MW01617808

GIS for Property Tax and Assessment Professionals

GIS for Property Tax and Assessment Professionals

Written by:
Margie M. Cusack
Paul E. Bidanset
Daniel J. Fasteen, PhD

IAAO PUBLICATIONS

An imprint of the International Association of Assessing Officers

GLOBAL HEADQUARTERS

314 West 10th Street | Kansas City, Missouri 64105

PHONE AND ONLINE

816-701-8100 | info@iaao.org | www.iaao.org

PROPERTY

COPYRIGHT AND LCN

Knowledge Area 1: Working with the Legal Framework
GIS for Property Tax and Assessment Professionals

Library of Congress Control Number: 2018953548
ISBN-13: 978-0-88329-243-3 (Paberback)
ISBN-13: 978-0-88329-244-0 (EPUB)

10 9 8 7 6 5 4 3 2 1

Published in the United States of America

First Edition | First Printing

BOOK DESIGN AND LAYOUT

Pop! Creative Design | popdesigned.com

Table of Contents

Acknowledgements

Thank you to the IAAO Executive Board for their ongoing support. Without their backing, this book would not have been possible.

Very special thanks go out to Brent Jones (Global Manager, Land Records/Cadastre, Esri) and Wendy Nelson (Executive Director, URISA) for their support and continued guidance, and to The Appraisers Research Foundation for providing IAAO with a TARF grant that was used to research and develop spatial valuation content in Chapter 5.

We also are very grateful to those who provided case studies for this book:

Neal Carpenter
President, CEO
The Sidwell Company

Peadar Davis, Phd, MRICS
Ulster University,
Belfast, Northern Ireland

Jennifer Kuntz
CycloMedia Technology, Inc.

Randy Lahr, SAMA
Senior Appraiser
Stearns County Assessor's Office

Michael Lomax
Director, Assessment
Esri Canada

Nate Maher
Appraiser Supervisor - Desktop Review
Maricopa County Assessor's Office

Michael McCord, PhD
Ulster University
Belfast, Northern Ireland

William McCluskey
African Tax Institute
University of Pretoria
Pretoria, South Africa

Pat O'Connor
President
O'Connor Consulting Inc.

Carmela Quintos, PhD, MAI
Assistant Commissioner, Property
Valuation & Mapping
Department of Finance
City of New York

Matt Sorensen
Sr. Vice President
Midland GIS Solutions

James H. Thomas
Assessor
City of South Portland, Maine

John Watterson, GISP
Martin County Property
Appraiser's Office, Florida

Russ Wetzel
GIS Analyst
Midland GIS Solutions

Scott Zubek, GISP
Director
Tioga County GIS Department

Preface

The International Association of Assessment Officers (IAAO) is honored to publish this guidebook—the first in our industry specifically tailored to describe the multiple uses of geographic information systems (GIS) tools in property tax valuation units throughout the world.

Written by property tax professionals, the following pages contain valuable information about the hands-on uses of GIS applications and technologies within all aspects of government valuation procedures and administration. GIS has become an indispensable tool for valuation professionals as it provides a platform for data to be easily displayed and facilitates the discovery and analysis of new patterns and influences on property value.

We hope this guide will strengthen your understanding of the role of GIS; help you implement improvements in all phases of land and property tax administration; and spark your imagination to envision creative future uses of GIS that will further enhance uniformity, fairness, and transparency in government.

Dorothy Jacks, CFA, AAS
Palm Beach County Property Appraiser
IAAO President - 2018

Introduction

Who We Are: We Value the World

The International Association of Assessing Officers (IAAO) is a nonprofit, educational organization founded in 1934. Its mission is to promote innovation and excellence in property appraisal and property tax policy and administration through professional development, education, research, and technical assistance. Its 7,700 members are government officials and others interested in the administration of assessment and property tax. All IAAO members subscribe to IAAO's Code of Ethics and Standards of Professional Practice.

IAAO is the primary publisher, educator, and leader of standards in the field of mass appraisal and assessment administration. As a standard-setting organization, IAAO has published standards aimed at improving all aspects of assessment practices. As an educator, IAAO has established a curriculum of 30 courses and 28 workshops, including courses concentrating on GIS through which valuation professionals can supplement their professional training, and individuals interested in pursuing a career in land and property tax administration can enhance their education. We offer the only comprehensive program of mass appraisal courses in the world. In addition, we offer special seminars and an annual international conference on assessment administration.

IAAO recognizes that assessment administration is a specialty within public service and that assessment personnel are relatively mobile. We therefore offer professional designation programs, including GIS-oriented designations, to certify the competence of individual assessors and to attest to this competence when career paths cross state/provincial lines.

For more than 80 years, IAAO has provided objective, insightful advice to assessment jurisdictions around the world. Promoting innovation and excellence in property appraisal and tax policy, IAAO continues to promote innovative solutions and offer technical assistance for the assessment industry worldwide.

Why Did We Write This Book?

Over and over again, we hear the same question from assessment and valuation professionals from all across the world: Will **geographic information systems** (GIS) technologies *really* help improve my operations and valuation processes? And the answer is a resounding YES. In fact, over the next 10 years GIS will completely transform valuation processes and how we share information globally.

This book was created to provide an easy-to-use, step-by-step guide for valuation officials on how GIS technologies can be used within government offices to improve efficiency, accuracy, and transparency in order to better serve the public.

This book is not a comprehensive textbook that describes all the mathematical components enabling this technology, nor is it an academic review of literature regarding the vast world of GIS. Instead, it is a concise and targeted compilation of where and how to implement GIS technologies in order to most effectively improve the processes and operations of valuation offices. It is, in effect, a collection of best practices that can be applied in any valuation office, anywhere in the world.

What is Our Goal?

Our goal is to demystify the many uses of GIS for assessment professionals, and to give the reader targeted information so that they can make educated decisions about which GIS technologies will be most useful to implement within their operations. The ultimate goal for any newly implemented technology should be to improve valuation development and its defense, and increase operational efficiency and taxpayer access to information.

- **The book starts off with Chapter 1, GIS and the Valuation Official,** which explains what GIS is and how it relates to valuation agencies and the official duties of government valuation administrators. Chapter 1 suggests that small jurisdictions would benefit from platform sharing, and concludes with a brief discussion about preparing for the implementation of a new GIS.

- **Chapter 2, Land Record Management in GIS,** focuses on the use of GIS in the management of land and property records. Property tax maps and property recording documents (such as legal descriptions of mapped property lines) are addressed, as are the challenges that assessment officials face during the maintenance of these property tax maps over time. Property tax maps can have varying levels of accuracy, as discussed in this chapter. We will also cover the development of commonly used data sets that will integrate easily with a GIS.

- Chapter 3, **Desktop Measurement, Data Collection, and Verification, moves past Chapter 2's discussion of the creation of a land record management** system within GIS, and on to the use of the newest GIS technologies. These technologies provide the ability to create electronic sketches on a desktop and come with a variety of electronic measurement applications for use on different properties. These tools facilitate accurately-drawn land dimensions (which define a property's boundaries) and building improvement dimensions.

  This chapter also describes new mobile and electronic field data collection technologies, including how they work and how they integrate into the GIS, thereby enhancing captured data characteristics for the development of property value. Equal attention is given to both established GIS technologies (such as **oblique aerial imagery** and **street-level imagery**) and to the newest GIS capabilities (such as LiDAR, 3-D imagery, and the use of drones).

  All of the GIS applications described in this chapter help to maintain accurate property records and facilitate change detection in property tax maps over time. **Change detection** techniques and GIS **desktop review** procedures are explored and applied to routine assessment operations and government property tax systems.

- Chapter 4, **Value Mapping and Geographical Analysis**, will explore the use of GIS technologies for these purposes, including how GIS allows analysts to "**geovisualize**"—that is, it gives them the ability to project any data being used to develop property values onto a thematic map. Value mapping assists various duties within the valuer's office, making many job functions easier to perform.

IAAO standards, which guide agencies seeking to achieve or improve the fairness and equity of their valuation estimates, are easier to meet when geographical analysis is used. Chapter 4 discusses how to use GIS through geovisualization to improve both horizontal and vertical equity. It also provides a framework for navigating some of the most common ways that value mapping and geographical analysis are used in the development of property values, including statistical testing, quality assurance and control procedures, and the preparation of a valuation for public appeal or review by oversight agencies. The future uses of value mapping and geographical analysis are enormous and will be revisited in Chapter 7.

- Chapter 5, **GIS and Mass Appraisal Models**, describes the relationship between GIS and mass appraisal modeling. A brief description of **spatial modeling** serves as the primer for this chapter. Specific types of spatial modeling techniques are then addressed and explained, which range from the appropriateness of different spatial variables to more advanced spatial modeling techniques, such as **geographically weighted regression (GWR)** and **response surface analysis (RSA)**. The benefits gained by using these techniques (namely, increases in model accuracy) are described in detail.

  Most importantly, this chapter hopes to answer questions that valuation professionals routinely ask regarding how they can improve their current valuation mass appraisal models through the application of GIS technologies. This chapter includes creative and innovative modeling techniques taught and shared by IAAO, and, although this chapter might be considered to be fairly technical—even by the standards of an industry professional—it is valuable and transferable information for personnel who are responsible for the continued improvement of their office's mass appraisal and spatial modeling.

- Chapter 6, **Using GIS to Serve the Taxpayer**, turns the focus toward the three contingencies that we as valuation professionals most serve: taxpayers, the public, and other government agencies. A government GIS mapping system is the framework upon which information and data relevant to the public are displayed. This chapter describes the many ways that maps can be used to help valuation offices illustrate how values are developed. Special attention is given to the emerging use of mapping applications for assessment-related functions, such as in the analysis of property appeals, mapping homeowners and other property exemptions, or displaying comparable properties on a map.

  This chapter also explores the use of open data portals within the framework of GIS, inter-governmental sharing and cooperation, using GIS for tax policy analysis, and the development of integrated property tax systems. **Interactive GIS tools** for taxpayers' use and the most current advances in web-based map creation and value analysis are touched upon as well.

- Chapter 7, **The Future of GIS**, looks at how emerging technology trends (including the latest GIS technologies) are game changers for government valuation offices and assessment administration. New mapping applications for valuation development and review are being created every day, and new virtual assessment tools could become a mainstay in future years and assist countries that are beginning to develop new property tax administration systems. Another newly available tool is the colossal amount of open data now available, which drive the increased data-sharing initiatives within valuation agencies throughout the world.

Chapter 7 also reviews ongoing, global efforts to standardize data related to property tax. IAAO particularly supports efforts made to standardize mapping techniques and land and building data. Finally, the chapter concludes with a discussion of the role that IAAO has taken in improving land administration and property tax systems worldwide.

Since these chapters are driven by topic areas that are found within routine government valuation operations, chapters can be handled individually—that is, a specific chapter can be skipped over if it doesn't apply to you, or another chapter handed off to the employees in your office who perform the specific duties it describes.

Each chapter contains real-world examples that tell a brief story about the practical application of GIS tools and technologies within an innovative valuation office. Relevant GIS terms are also defined within each chapter.

The **appendix** includes GIS resources that were specifically chosen to give valuation officials further assistance with the development of GIS technology.

Important Definitions

Some of the terms used within the assessing profession vary widely both internationally and across the industry. We have therefore decided to include a short glossary in this introduction, to define those words that tend to have the most inconsistent usage throughout the business. The following list explains how we define certain terms in this book. Words that can be used interchangeably to mean the same thing are separated by slashes.

- **Assessor/Valuation Officer**
    1) The head of an assessment agency, sometimes used collectively to refer to all administrators of the assessment function.
    2) The public officer or member of a public body whose duty it is to make the original assessment or determine the value of a property for the purposes of property tax administration.

- **Appraiser/Valuer/Analyst/Specialist**
    1) One who estimates the value of property.
    2) More commonly, one of a group of professionally-skilled persons responsible for the development of property valuation.

- **Automated Valuation Model (AVM)/Computer-Assisted Mass Appraisal (CAMA) Model/Hedonic Pricing Model**
    1) A statistical algorithm that estimates market value and/or value adjustments based property and market characteristics. When used for property tax purposes, this model is typically regression-based, and coefficients serve as value adjustments.

- **Reassessment**
    1) Most commonly, the revaluations of all real property by the regularly constituted assessing/valuation authority, as distinguished from assessment on the basis of valuation, most or all of which were established in some prior year.
    2) The relisting and revaluation of all property, or all property of a given class, within an assessment district by order of an authorized officer or body after a finding by such an officer or body that the original assessment is too faulty for correction through the usual procedures of review and equalization.

- **Revaluation/Recalibration**
    1) The mass appraisal of all property within an assessment jurisdiction accomplished within or at the beginning of a reappraisal cycle.

CHAPTER 1
GIS and the Valuation Official

1 | Introduction

In 1968, Roger Tomlinson—the "father of GIS"—introduced the term "geographic information system" in his paper "A Geographic Information System for Regional Planning." In the decades since, GIS has grown and developed past what many people in the 1960s could have possibly imagined. Certainly, it has improved the ease, efficiency, and accuracy of many of the tasks we perform in the assessment profession. Throughout this book we will discuss the ways in which GIS accomplishes these feats, and how valuation officers can best take advantage of all the tools that GIS has to offer. But before we get down to the nitty-gritty details, let's start by defining what exactly a geographic information system is.

Geographical information systems, or GIS, are electronic software systems that enable the creation of electronic maps that are geographically coded to allow for the addition of supplementary data layers. In other words, data that are captured and stored by GIS technologies can then be incorporated into a map to create a geographical display of that information. GIS software enables the management, analysis, and flexible presentation of data. Any datum that has some type of locational identifier can be stored in a GIS and represented on a map.

Since location is such a critical component to value, GIS is an extremely powerful tool in the discovery of new patterns and influences on estimates of property value. Using GIS tools to map and display information allows government agencies to quickly visualize patterns and provide facts to the public in a way that is easy for them to decipher. As they say, a picture is worth a thousand words.

Figure 1.1 contains a diagram of the layers in a GIS that are typically utilized by assessors. See Chapter 2 for a more thorough explanation of how layers are developed in GIS.

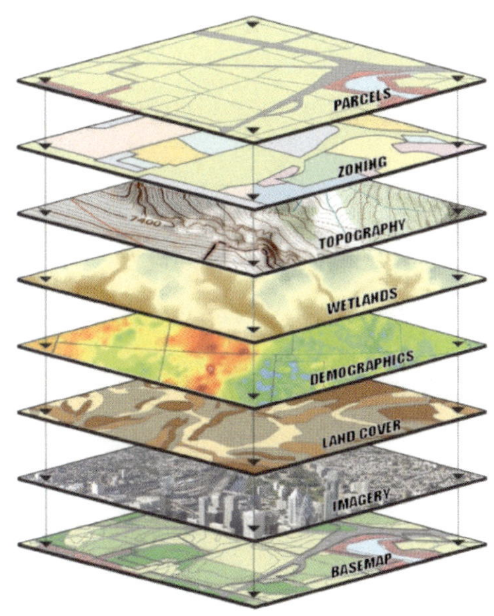

1.1: Introductory diagram of the typical layers of a GIS platform utilized by assessors
Source: Esri and its licensors

2 | Why GIS?

Since the valuation industry works closely with property records that consist primarily of land and building information (and relies upon the accuracy of those records), meticulous electronic mapping systems are critical in enabling the efficient operation of the many tasks performed by a valuation office. GIS facilitates the ability to create and maintain accurate property records for the development of fair and equitable property taxation over time. Newer GIS technologies have even ushered in the creation of new value indicators that enhance value prediction in the assessment industry. Through GIS, targeted locational attributes have been discovered that can be added to existing valuation mass appraisal models to improve accuracy at a very detailed level.

GIS provides a key software platform for government agencies worldwide, especially since a large majority of government services are tied to defined land boundaries. For valuation officials, almost every duty or assigned agency function is defined by property borders. GIS software provides a framework to support and boost the performance of common valuation tasks, functions, and operations, including:

- Measuring and listing accurate property identification contained in property tax records
- Maintaining accurate property descriptions and identification of property tax records, including keeping track of any changes that occur
- Creating accurate property tax maps
- Detecting and tracking new property construction and property permits in order to capture additional property value
- Collecting new data in the field through electronic devices and software that develop efficient routes for field personnel
- Performing electronic desktop review and data verification
- Enabling the spatial observance of data patterns for increased valuation awareness (patterns which would have most likely been missed if the data had been housed outside of GIS platforms—say, in a tabular format)
- Creating new geospatial mass appraisal modeling methods that can fine tune locational influences on property value at a very localized level
- Sharing data among government agencies
- Measuring the accuracy, uniformity, and equity of property value estimates through maps that allow for valuation predictions to be reviewed spatially
- Displaying the official listing of property values for taxation purposes
- Improving the processing of taxpayer appeals of property values
- Administering and maintaining taxpayer and property exemptions records, as defined by law
- Reporting property values to oversight agencies
- Improving taxpayer awareness and providing information to the public
- Improving online government services and interactions with taxpayers

These common core tasks and operations are performed by all valuation offices—no matter how offices differ as defined by administrative rule, common practice, or law—and are supported and strengthened through the use of GIS technologies.

It should be noted that all of the common valuation procedures and tasks described in the preceding list are addressed within IAAO standards. IAAO standards detail best practices for the government agencies responsible for the development of land administration and property tax valuation. These documents can be used as your guidepost as you move forward with new technology initiatives, and can be found—along with all other resources provided by IAAO—at **www.iaao.org.**

Using GIS applications not only improves office outcomes by facilitating operational efficiencies, but also bolsters the achievement of the most important elements of a property taxation system as described by IAAO standards: uniformity, fairness, and equity.

3 | Getting Started

It takes time to prepare for and implement a new GIS, or even to add technology advancements onto a current system. It is strongly recommended that a valuation agency take a multi-agency or multi-jurisdictional approach to GIS development, rather than thinking of it as a single government agency solution. In most government structures, anything a valuation agency wishes to create in GIS to improve their efficiency or values could likely be used in some fashion by another agency as well.

As stated, GIS is essentially an electronic mapping system with multiple layers of data that can be shared and analyzed in many versatile ways, so government collaboration is key when developing a meaningful multi-agency system. Security can be set up so that the agency responsible for specific data elements is designated as the only agency authorized to change or manipulate those data, alleviating safekeeping and protection concerns. GIS platforms are designed to be data-sharing platforms, and keeping this fact in mind will help a jurisdiction find the most optimal GIS solution possible for an entire government unit, or even across multiple government units.

3.1 | Advice from a Pioneer

City of Norfolk assessor and former IAAO president, William A. "Pete" Rodda, CAE, RES, has been something of a pioneer in GIS development for governments and valuation offices. "In the past, GIS usually started in a single department within a local government. Over time other users came on board, but the environment tended to be restrictive. Copying files to a data warehouse server promoted the sharing of data but did not adequately address latency and management questions."

In the 1990s, Mr. Rodda reached the conclusion that the assessor's data are the "center of the GIS universe." The people who compose the clear majority of GIS users need to know: (1) parcel ownership, (2) parcel location, (3) parcel size/shape, (4) what improvements are on the parcel, and (5) what the parcel is worth. Because these five criteria form the foundation upon which so much other information is built, Mr. Rodda believes it is imperative that assessment operations are able to address these questions quickly, accurately, and consistently.

Use GIS to fully leverage the information and data that are already available from a multitude of government and jurisdictional agencies.

"The sharing of real-time information in GIS allows for the further leveraging of data gathered by multiple government agencies or jurisdictions, both from the activities they are already performing and data that have already been accumulated. The use of GIS in this manner leads to more informed decision-making, faster and more effective government responses, and more efficient use of human and financial resources. Government agencies are already paying for the accumulation of data that can be easily placed into a GIS. Information such as the property owner's name and address, the parcel identification number, property address, tenant's name and address, parcel-specific data, and any other data you can think of could be added as new data layers to the GIS technology solutions that are being utilized by other government agencies."

As Mr. Rodda further describes, "multiple government agencies can benefit from any future GIS technology the valuation office might be considering. For example, if the assessor's office wishes to implement an electronic desktop measurement technology, sharing that software tool—or the data verification that results from that software tool—is a sensible thing to do. The collective benefits gained by multiple agencies sharing a single new technology solution should be thoroughly discussed with other government agencies and jurisdictions. Why? To ensure that the format and objectives of that shared tool are consistent with the goals and objectives of the local government, and that the software and data output are compatible with other software systems currently in use.

"Multiple government agencies should join as collaborative advocates (partners) for GIS technology development. Partnerships would include (but are not limited to) the assessor, the planning director, law enforcement and emergency service officials, the inspections director, and the IT director. Ultimately, government executive leadership will come to understand the efficiencies and benefits that can be accomplished through further GIS technology."

Here are some additional tips from Mr. Rodda on how to get started.

3.1.1 | Involve Management Early and Often

Management should be involved early and often in the process of selecting a new GIS technology. This includes elected officials and government department managers. It is extremely important to involve management because they are the ones who make decisions regarding the budget, and they will be the ones who ultimately decide whether the GIS project will be funded. The assessor's office, as an advocate for GIS, should help facilitate the crossing of historic jurisdictional and department boundaries by explaining the benefits gained by sharing data, while also highlighting the ability that users have to absolutely maintain jurisdictional control of data within GIS.

Promoting cooperation and collaboration is the key to the successful implement and use of GIS. Particular emphasis should be placed on the added government transparency that GIS provides to taxpayers and on the cost-effectiveness of GIS software solutions as they increase efficiency and enhance operations.

3.1.2 | Answer Question of Funding

Questions need to be answered and justifications made regarding GIS funding. It is important that you are prepared to answer how GIS upgrades and technology solutions will be paid for. Consider finding out what is being paid for existing software and hardware that are used for GIS purposes in various government departments. It is entirely possible to leverage existing software licenses and equipment by consolidating them into a more efficient, online, web-enabled GIS. New GIS solutions save money on current GIS costs across government agencies because new technologies in GIS are often less expensive.

Another question to ask yourself is whether existing staff members have their current duties streamlined and made more efficient by GIS. Redundant or duplicate government tasks can be consolidated or eliminated by GIS, which enables personnel to perform other analytical tasks and improve overall government performance. Some of the tasks that are performed in a valuation office that could be streamlined by new GIS technologies and data consolidation include:

- Some data entry tasks
- The deployment of personnel to the field in order to verify property characteristics
- Some research tasks
- Information requests from other government agencies

A final consideration surrounding the question of how GIS will be funded is the potential revenue created from the sale of products that can be generated using the GIS. Maps and lists of properties that contain specific characteristics can easily be generated and sold to the private sector. For example, what would a map and list of property owners who own swimming pools be worth to a pool maintenance company located in a county in Florida? Another example is that information regarding sales and other trends can be included in the sale of customized products for real estate professionals. It should be noted that some information within the GIS falls within public records laws across the world, but the generation and sale of customized products should still be considered.

Sharing a GIS with other government jurisdictions through governmental agreements will also reduce the cost of this technology. Consideration should be given to developing these partnerships in order to reduce the costs associated with GIS development for both jurisdictions, and to enhance further data sharing in the future.

3.1.3 | Become an Agent of Change

GIS symbolizes the new world of technology and data sharing that is happening not only in government but also across the globe in private industry. As an advocate for GIS technologies, the valuation office will have to become a change agent. Convincing other government officials to share information outside of their agency and change their entrenched operational processes can be a challenge. One very important consideration to remind them of is that sharing GIS data will augment and improve information for their agency as well.

Assuring government officials that current data will be enhanced and augmented through GIS data-sharing is one of the initial challenges a GIS technology leader in

As with any new technology initiative, fear is a factor that will need to be addressed and overcome. Fear can be alleviated through clear communication with government executives and other leaders regarding the collective benefits of the project.

---

government will face. Another challenge will be convincing officials that new technology solutions are as safe as locally maintained data, and have the same protections in place. An IT expert should champion the answers to these important questions, using their expertise and clear explanations of security software innovations within the government sphere to assuage any fears and allow officials to move toward new GIS platforms confidently.

It is just as critical to prepare for the changes that a new GIS technology will bring, including greater transparency and interaction with the public. For example, if a new GIS technology allows a taxpayer to pick and choose their own comparable properties, these new tools will allow them to discover things that they may have follow-up questions about. It is very important to discuss and prepare for these anticipated changes in the way the public is able to interact with government agencies so that these agencies are not caught off guard and are ready to provide quick responses.

3.1.4 | Learn from Others

Throughout this entire process, be sure to learn from others who have already completed GIS implementations. Their experiences and lessons learned will be very helpful as you plan your own implementation project. Develop a GIS steering committee to keep key players within government entrenched in the maintenance and further development of a GIS. Having a GIS information office or department as a coordinating agency helps to keep on-site GIS expertise and project management consistent, coordinated, and cost-effective.

In summary, as with all things in government, the key to a successful GIS project is proper planning and communication. This includes creating a simple description of GIS efficiencies and cost-saving benefits for the public.

3.2 | Utilize Industry Best Practices in the Creation of GIS

As a starting point, it is very helpful to become acquainted with the IAAO *Standard on Digital Cadastral Maps and Parcel Identifiers.* Reading this standard will familiarize you with the words and jargon that are often used to describe GIS. The standard covers the core components of a digital cadastral mapping system, including details about the following aspects:

- Preparing to create a mapping system
- System maintenance
- Quality control
- Property or parcel identification
- Maintaining parcels

The standard will also provide you with a foundation for justifying the expense of creating new GIS solutions, since this standard represents the valuation industry's consensus on how to create a "best practice" GIS that is specifically tailored for offices related to valuation and land administration for property tax systems.

3.3 | Create a Scope of Work

At the start of a GIS project, it is always beneficial to write down the scope and the objective of the project. This exercise produces an executive summary of a GIS project not only because it provides a document to help other government officials understand what it is you want to accomplish, but because it can be used as the starting point for the development of a statement of work.

Another good strategy is to obtain the assistance of both IT and GIS experts who will act as the office's advocate for this project. Many government valuation agencies already have some type of GIS in place. If you wish to augment an existing GIS with new GIS technologies to improve valuation practices and procedures, research should be done to make sure that any new GIS applications are easily compatible and will integrate with the existing GIS. If ensuring compatibility requires extensive resources, a determination needs to be made as to whether a government agency should consider a newer, more updated GIS solution altogether.

Specifications for new GIS implementations can be technical in nature, so government officials should make use of personnel who possess the mechanical knowledge to decipher the capacity of a proposed GIS software solution. They will have the expertise needed to determine whether the proposed solution meets all technical requirements as defined by the entire jurisdiction.

A cost comparison against similar software solutions and applications should also be conducted and analyzed. The costs associated with a GIS technology solution are not limited to the purchase of software, but also include the additional cost of project implementation and any training required to bring agency personnel up to speed. Valuation officials need to do their due diligence to make sure that the entire cost of all phases of a technology upgrade are taken into account, and that project documents are thoroughly developed and reviewed by key subject matter experts and intended users of the software before embarking on any new project.

A statement of work is a document that is often prepared by government agencies to describe in detail the scope and objective of a proposed technology upgrade. A statement of work usually requests an analysis of current processes (also called processes "as is") within the office operations to serve as a starting point. It often includes a "to be" process as a second development phase. This helps define not only what technology components must be included in the new installation, but also identifies what operational procedures in the office will be affected and ultimately changed due to this technology solution.

Knowing what your starting point is and where you want to be at the end of the GIS project is critical in analyzing the operational changes and efficiencies that are expected to result from this project.

Many GIS technology advancements involve a chosen private industry solution and the provider will assign a project manager to oversee and run the project. Each government agency involved in the project should assign a project manager (or subject matter expert) to serve as the advocate for their office. Although each of these people will come from a different agency, they should also have equal standing and their purpose is to make sure that all expectations, specified outcomes, and deadlines are being met throughout the installation. An agreed-upon project timeline will often include signoff points—sometimes called "milestones"—in the project's development to ensure that it is progressing at the expected pace.

Some projects also involve data conversion efforts, meaning that existing data must be reworked into a format that a new solution will accept. Again, IT experts should be called upon to assist in defining this critical phase of the project. All GIS technology upgrades need to have time allocated for user testing and user acceptance testing (UAT). Including time for user testing and approval by the government agency is imperative to a successful GIS project. Staff training is also important to ensure that they understand how to use the new product.

Maintenance and support agreements for new GIS solutions are a reoccurring cost for any technology solution once the initial implementation has been completed. Researching the cost of maintenance agreements for proposed GIS solutions can be accomplished by asking comparative valuation jurisdictions what they pay for theirs. Having this comparison is often very helpful, as it provides documented support for proposed maintenance cost estimates. Lessons can also be learned from colleagues who are either investigating similar technology upgrades or who have completed similar GIS projects. These can be used for justifying a proposed budget.

Many government agencies have formalized procedures that include requests for proposals (RFPs)/considerations to initiate technology projects. Any proposals that are received can be compared against one another. As stated earlier, IT and GIS experts can act as your agency's advocates to make sure that system requirements within the RFP thoroughly describe all of your agency's needs. Having clearly stated expectations and outcomes that are understood and agreed to by both your government agency and the solution provider are key to successful and timely projects that do not go over budget.

If you are an official who is investigating the creation of a new GIS that includes the conversion of paper maps to electronic maps, the technical requirements and digital map conversion costs for this implementation should be thoroughly investigated.

Many countries are just beginning to develop land administration systems for the future implementation of a property tax system in the hopes of one day providing expanded government services. Again, preparing and planning for large GIS initiatives requires detailed specifications and phased planning for all stages of the project's development, from start to finish.

3.4 | Being Cost-Effective

The cost of newer GIS solutions is certainly far less than what agencies had to pay in the past. Newer GIS technologies are not only cost-effective, but also offer functions and applications that were not available even five years ago. Additionally, web-based GIS solu-

GIS technologies offer the most cost-effective approach for bringing new innovations, efficiencies, and open data into the operations of a government valuation office.

---

tions can provide easy access to large amounts of open data, which can be used to enhance valuation development. Chapter 6 will discuss the critical role that GIS play within the newly emerging realm of integrated property tax systems.

4 | Funding for GIS and Technology Advancements

As with all computer software systems, GIS requires periodic upgrades and new application enhancements, and (as mentioned) maintenance and support agreements. If it is at all possible, a dedicated revenue source should be developed to pay for GIS maintenance and upgrades.

This dedicated revenue source could be a line item, meaning that a specific portion of the government's annual budget is allocated for GIS maintenance and development. Another option is to create a new revenue source through governing law or ordinances. For example, an additional fee could be charged for all transfers of property ownership that are recorded in a registry office for the specific purpose of maintaining and further developing that government's GIS. Other fees related to land records or the customized sale of GIS products throughout government agencies could also be considered as potential sources when establishing funding to maintain and support a land records management system.

4.1 | Considerations for Training

As mentioned, training should be included as part of any GIS technology upgrade. Strategic training involves identifying a couple of people within an organization who are willing to take on the role of subject matter experts and trainers in a new GIS software solution. Depending on the size of a government agency, providing training to one or more "power users" first is a proven approach to success, since not only will a power user become very familiar in how to use the software, but they also understand the office's guidelines and procedures. New technologies can then become an integrated part of improved standard operating procedures.

Providing funding for continued GIS training for all office personnel is key. Educating personnel will ensure the quick application of new GIS technologies within an agency. A Cadastral Mapping Specialist (CMS) designation for valuation specialists is offered by IAAO to those who would like to master the use of GIS. Links to information regarding all IAAO professional designations, including a new IAAO designation, the Mass Appraisal Specialist (MAS), are found in the appendix and at **www.iaao.org.**

Online opportunities for GIS training are enormous and sometimes overwhelming. IAAO offers very targeted training for valuation personnel in the use of GIS in all aspects that are necessary to the assessment profession. Urban and Regional Information Systems Association and Esri also offer GIS training for land and property tax administration officials.

Online social media GIS user groups and free resources are also in abundance. The appendix offers a starting point of suggestions for GIS training and educational materials for valuation professionals.

5 | Conclusion

GIS mapping tools are extremely powerful and can enhance and increase efficiency for a large variety of functions performed by valuation offices, and by every other agency involved in land and property tax administration. Chapters 2 through 6 will focus on how to optimize office procedures using GIS.

The implementation of GIS technology can take some time to properly prepare for, but these efforts are well-rewarded by an increase in government transparency and public information delivery, since mapping real estate valuation data makes it easier to understand. GIS technology will improve valuation development by augmenting existing information, and will allow for easy-to-use online taxpayer interaction tools—something that will soon be commonplace.

CHAPTER 2
Land Record Mapping in GIS

1 | Introduction

Ownership of real property in the United States and around the world is a coveted privilege that many people strive to attain. In the United States, many Americans equate owning a home with the American dream.

There are several processes that go into land and property acquisition and dividing land into parcels. A **parcel** or **cadastre** is a unit of land defined by a series of measured straight or curved lines that connect to form a polygon. Land parcels are only a small part of an infrastructure utilized to maintain records on land-related policies and the management of those policies. This infrastructure—referred to as **land administration**—includes property records, parcel identification, deeded legal descriptions, valuation information, and various other pieces of recorded information pertaining to the land and property. *Figure 2.1* shows how the components of the land administration system are integrated.

The property valuation process is well-integrated into this system and is heavily influenced by surrounding geographic supply and demand forces that shape the landscape. Thus, governments who maintain a good land administration system often have the resulting advantage of a good mass valuation system. The maintenance of complete and accurate land records data, cadastral identification, and cadastral boundaries are essential to the valuation process. These elements provide the proper foundation for visualizing and analyzing both property data and property valuation data.

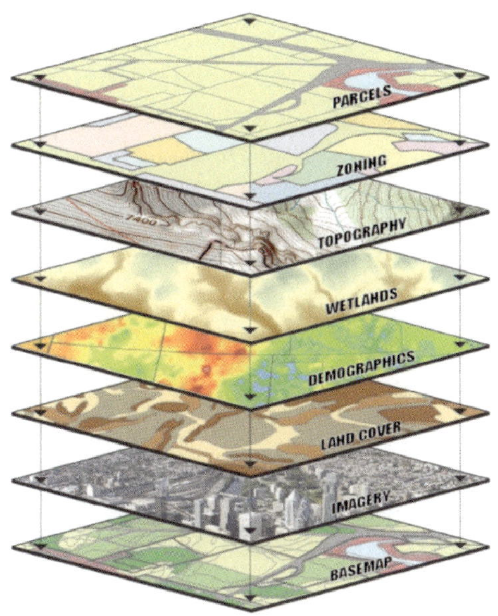

Figure 2.1: Land Administration Layers.
Source: Esri and its licensors

This chapter will describe cadastral mapping and land administration as it relates to the valuation process. We will dive into the IAAO standards in order to develop a "best practices" perspective of the elements involved in maintaining land records and cadastral information. Next, we will discuss some best practices for creating and displaying cadastral data through both a static medium (paper maps) and a dynamic medium (web-mapping applications).

This chapter will also provide a discussion of land parcel maintenance and deeded boundary line adjustments, including an explanation of their importance within the valuation process. Furthermore, we will briefly touch on land records data and data collection, which will be the primary focus of Chapter 3. There are a number of different roles that an assessor might assume in maintaining all or part of a land administration system, and this chapter provides a general overview of common maintenance methods that are utilized within the industry.

2 | Cadastral Identification and IAAO Standards

The assessor's role in maintaining cadastral information stems from the general collection of property data. In order to list and value property, the assessor must first identify the absolute location of the property, which might include the property's address or geolocation coordinates such as latitude and longitude. These location identifiers are used to display the location of parcels on maps, as they typically reflect the actual geographic coordinates of the center of the land parcel. The center of a parcel's geometry is referred to as the parcel **centroid**. This centroid is often assigned a unique parcel identifier at the time of the parcel's creation, and is only changed when or if the boundaries of the parcel change. It is a best practice to create new parcel identification numbers when such a change occurs—rather than reusing the old ones—as the careful maintenance of the lineage of a parcel's geometry and related records are incredibly beneficial when an audit is conducted.

IAAO has identified several best practices for the administration of parcel identification that greatly enhances the standardization of cadastral data. A parcel identification number, often referred to as a PIN, is a unique number assigned to each individual parcel to denote its existence within the cadastral system. This number will often be utilized and referenced within land administration to denote land and property data, valuation data, recorded documents, or other information specific to the legal description of a particular property. This number is also frequently referenced on maps in order to join property information within CAMA, and also in document management systems to link documents that pertain to those parcels.

2.1 | Characteristics of Parcel Identification Systems

There are six general characteristics that all PIN identifiers should possess, as indicated by the *Standard on Digital Cadastral Maps and Parcel Identifiers*. These same characteristics are described in greater depth in *Property Assessment Valuation*:

- **Compliance with standards** - All parcel identification systems should comply with all statutory policies set forth by the jurisdiction, regional, or state governments. This ensures compliance and standardization with other government entities.

- **Uniqueness** - This is the most important characteristic in a parcel identification system. A PIN must be able to distinguish one parcel from all the others. A PIN system should contain a single unique identifier to every parcel that includes the land, property, and everything attached to it. This ensures that it can be identified as a separate entity, and that other property data and records can be associated with that particular parcel.

- **Permanence** - The PIN should be permanently affixed to the parcel for the duration of its life cycle. A change should only be made if a boundary line adjustment or other transaction affects the geometry of a parcel.

- **Simplicity and ease of use** - The PIN system should be easy to use and understand. Documentation should be created that describes the meaning of each of the PIN's components, and how the PIN changes through property transactions and platting.

- **Ease of maintenance** - A process must be created to apply changes to parcel geometry when a deed, plat, or other instrument is recorded. Local governments are typically composed of several departments that have a contributory role in this process. It is essential that any maintenance performed on a PIN system is thoroughly documented, and that it is able to effectively accommodate changes to geometry and land records data, and also able to maintain the lineage or history of the parcel parents and children.

- **Flexibility** - The parcel numbering system must be flexible in order to accommodate various forms of boundary line transactions, including plats, replats, condominiums, air rights, easements, and leases. The numbering system should be able to accommodate simple boundary line adjustments, if needed, with enough spacing between digits to allow for new identifiers to be created in the instance that old identifiers are deemed obsolete.

A PARENT PARCEL is the original geometry of a parcel prior to its division into new parcels, and whose parcel identifier is often retired or given to one of the children.

A CHILD PARCEL is subdivided from the parent after land is divided through a legal description in a recorded document. New parcel identifiers are often created or descended from the parent. Land information is also subdivided from the parents to the children during this process.

LINEAGE is often maintained through CAMA systems or land records management systems as an auditing method for understanding how land ownership and parcel geometry have evolved over time.

2.2 | Types of Parcel Identification Systems

There are several different types of parcel identification systems used throughout the United States and internationally. Four of these types—geographic coordinate system

Latitude-Longitude -
"A reference system used to locate positions on the earth's surface. Distances east-west are measured with lines of **longitude** (also called meridians), which run north-south and converge at the North and South Poles. Distances north-south are measured with lines of **latitude** (also called parallels) which run east-west" (Wade & Sommer 2006). Identifiers are often labeled with "X" to denote longitude, and "Y" to denote latitude.

UTM -
A projected coordinate system that divides the world into 60 zones, each six degrees wide. Coordinates are typically given by the zone number and the easting and northing distance (in meters or feet) with the point of origin at the intersection of the equator and the zones central meridian. "X" is typically denoted as easting; "Y" is typically denoted as northing.

identifiers, rectangular survey system identifiers, map-based system identifiers, and name-related system identifiers—are called out by IAAO standards as best practices, since each of these types maintain a mixture of the six essential characteristics described in Section 2.1.

- **Geographic coordinate system identifiers** - A geographic coordinate system parcel identifier is based on the geolocation coordinates (often latitude and longitude, Universal Transverse Mercator (UTM) coordinates, or other coordinate system) of the centroid of a parcel polygon.

  Figure 2.2 depicts an example of a geographic coordinate system.

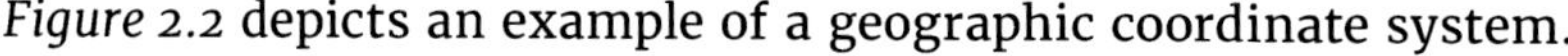

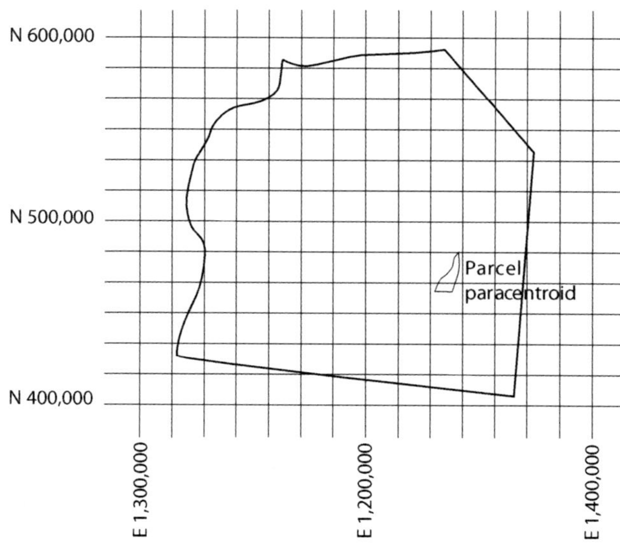

Figure 2.2: Geographic Coordinate System

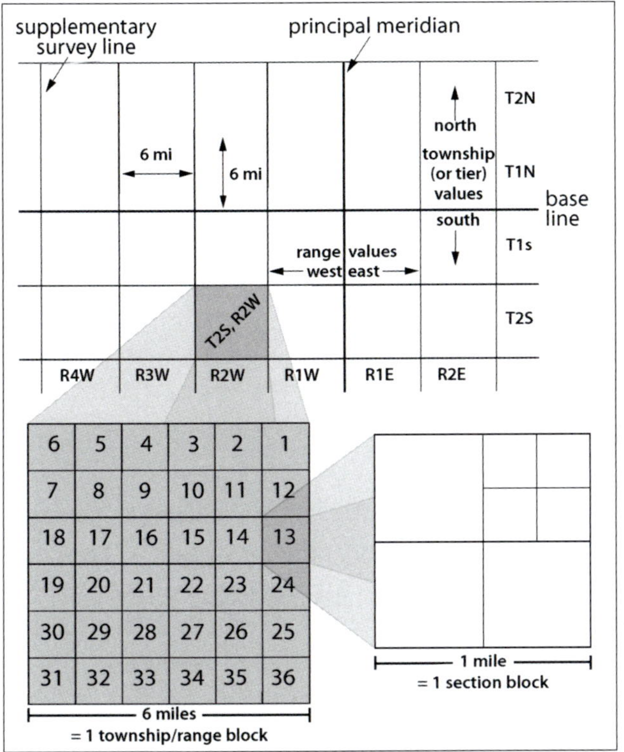

A SECTION is one-thirty-sixth of a township, bounded by parallels and meridians, equal to one square mile and containing 640 acres.

A TOWNSHIP is a quadrangle approximately six by six miles on each side, bounded by parallels and meridians containing 36 sections.

Figure 2.3: Rectangular Survey System

- **Rectangular survey system identifiers** - Rectangular systems are based on the Public Land Survey System (PLSS), in which identifiers are assigned based on the concept of a township, or a six-by-six-mile square tract of land. A parcel is often identified by the section, township, and range followed by the quarter section, and/or further subdivided by four 40-acre tracts of land. Platted parcels may also be based on the lot and block of a particular subdivision. See *Figure 2.3* for a sample rectangular survey system.

- **Map-based system identifiers** - Map-based systems utilize map books as a method of parcel identification based on the location of a block or group of parcels where that particular information is maintained. This method is not used much anymore, due to the rise of digital mapping systems and CAMA systems.

- **Name-related system identifiers** - This system uses the names of property owners to identify parcels. This system is rarely used and highly discouraged due to lack of conformity.

  Table 2.1 outlines some of the main advantages and disadvantages of each of these identification systems.

As noted in Section 2.1, additional information on these PIN types can be found in the *Standard on Digital Cadastral Maps and Parcel Identifiers*, and in the book *Property Assessment Valuation.*

Table 2.1: Advantages and Disadvantages of Various Identification Systems

TYPE OF SYSTEM	ADVANTAGES	DISADVANTAGES
Geographic Coordinate System	Unique Z-value for condominiums	PINS may be longer Permanence might be affected based on boundary changes
Rectangular Survey System	Preserves uniqueness and permanence	Not applicable to jurisdictions not using PLSS
Map-Based System	Easy to use and good for paper maps	Not good for digital or web maps
Name-Related System	Useful for paper maps	Not useful for digital or web maps Doesn't meet criteria of permanence, reference to geographic location, or ease of use.

Overall, the key element that an identification system must have is a consistent numbering system within a broad geographical area that is recognized by multiple levels of government organizations. The geographic coordinate system and rectangular survey systems are the more common reference systems used by assessor's offices, and they are also recommended by IAAO as the primary means for parcel identification. These systems form the foundation of land records management, cadastral mapping, and identification of land.

3 | Mapping Parcels for Valuation Purposes

The next phase of land administration, within the context of the assessor, is to map parcel boundaries and affix data to parcel geometry to provide efficiency for assessment staff in listing and valuing property. It is essential to have a robust mapping program within the assessor's office that combines several elements to overlay with parcel boundaries. These elements can include variables that are used in mass valuation models to visualize quality assurance, or to look at the location of a subject property and its proximity to an externality such as a commercial development or industrial area.

Visualizing a parcel and its information is essential to the assessor in conducting the tasks of identifying, listing, and valuing properties. Assessors utilize a cadastral mapping system as a base to reference, visualize, and extract information on parcels and property attributes. A **cadastral map** is a map that shows the boundaries of land subdivisions for the purposes of describing and recording ownership. There are several elements of a cadastral map that make it particularly useful for assessors.

3.1 | Elements of Cadastral Mapping

According to the IAAO *Standard on Digital Cadastral Maps and Parcel Identifiers*, several of the elements of a good cadastral mapping system are also major aspects of a land administration system.

Assessor's offices have a wide range of responsibility in maintaining quality and consistent cadastral data. The core elements are briefly overviewed in the following paragraphs, and many of these features will be referenced and expanded upon further in other areas of this book.

- **Geodetic control network** - This network refers to a collection of reference points on the earth's surface where the coordinates are known with certainty. These are often called "benchmarks," and frequently act as the starting point for an accurate cadastral survey.

- **Imagery/remotely sensed data** - Imagery is used to underlay or supplement cadastral mapping systems, as it provides a base upon which parcel geometry can be compared to actual images of the property. IAAO recommends that imagery be collected at minimum every five years to remain up-to-date. There are several different forms of imagery, including oblique, aerial, and street level, and other frequently used multi-spectral products such as LiDAR. LiDAR, or Light Detection and Ranging, is often used to produce high-resolution digital terrain models (DTM) and/or three-dimensional overlays of bare land. LiDAR uses pulsating lasers to measure distances between itself and reflective surfaces to develop a depiction of the terrain below.

In property assessment, LiDAR can also be used to digitize and create a 3-D layout of the interior and exterior of properties, and is often supplemented with aerial flights or street-level imagery capture.

- **Cadastral map layers** - There are various core layers that can be used to supplement a cadastral mapping system to provide needed relevant information. These layers might include the public land survey boundaries, subdivision plat boundaries, easements, rights-of-way, and building footprints. Other supplemental physical layers might include hydrographic and geologic layers. There could also be political boundaries including county, city, and jurisdiction; taxing districts; and zoning. See *Table 2.2* for additional suggestions of potentially useful supplemental layers.

Table 2.2: Supplemental Map Layers

CADASTRAL BOUNDARIES	PHYSICAL	ECONOMIC	DEMOGRAPHIC
Section boundaries	Slope	Income	Age Ranges
Township/Range	Soil Type	Distance to Work	Gender
Benchmarks	Topography	Income Ranges	Population
Parcel lines and polygons	Lakes, Rivers, Streams	Households	
Utilities			
POLITICAL	**IMAGERY**	**REMOTE SENSING**	**POINTS OF INTEREST**
Municipal boundaries	Aerial	DTM	Building Footprints
School districts	Oblique	LiDAR Points	Fire and Police Stations
Census Blocks/Block-groups	Street Level	3-D	Hospitals
Roads	Historical Imagery	Topography	Schools

- **Parcel identifiers** - As discussed in Section 2, a good identification system is: compliant with standards, unique, permanent, simple and easy to use, easy to maintain, and flexible.

- **Linking spatial and non-spatial data** - The greatest and perhaps most useful element with respect to a cadastral mapping system is the ability to link or "join" other relevant data. This essential element can serve many purposes for identification, visualization, and analysis.

- Each of these elements supports the others, and together they form the base of what IAAO's *Standard on Digital Cadastral Maps and Parcel Identifiers* calls a **multi-purpose cadastral environment**. This is defined as a cadastral map that contains data elements to support the assessor in all critical functions of their duties, including identification, listing, and valuing property. The multipurpose cadastral environment includes not only these base layers to support data creation and identification, but will also form the foundation of data analysis functions for valuation (which will be explained later in Chapters 4 and 5). The ways in which a cadastral map is published and utilized within the assessor's office depends on the map's purpose and use within the assessor's office.

3.2 | Purposes of Assessor Cadastral Mapping

Cadastral mapping can take place in either a static or dynamic environment—both of which will be described in greater detail—depending on how it will be used within the assessor's office. Some potential uses include identifying parcel boundaries as they relate to the parcel's aerial image, identifying a parcel's characteristics and attributes, and/or visualizing property variables and comparing the parcel to those that surround it for consistency. There are several mediums through which parcels may be mapped, and the medium that is ultimately chosen will depend on the user's purpose and access needs.

Static maps are standalone (non-interactive) maps which can be printed or displayed on a mobile device, capable of being carried to meetings or into the field with the appraiser.

Used for...
- Finding a property relative to other properties, such as comparable sales
- Keeping track of properties during a reappraisal
- Displaying nearby attributes

Advantages:
- Useful if connectivity is not an issue
- Useful if a computer is not available for the intended audience

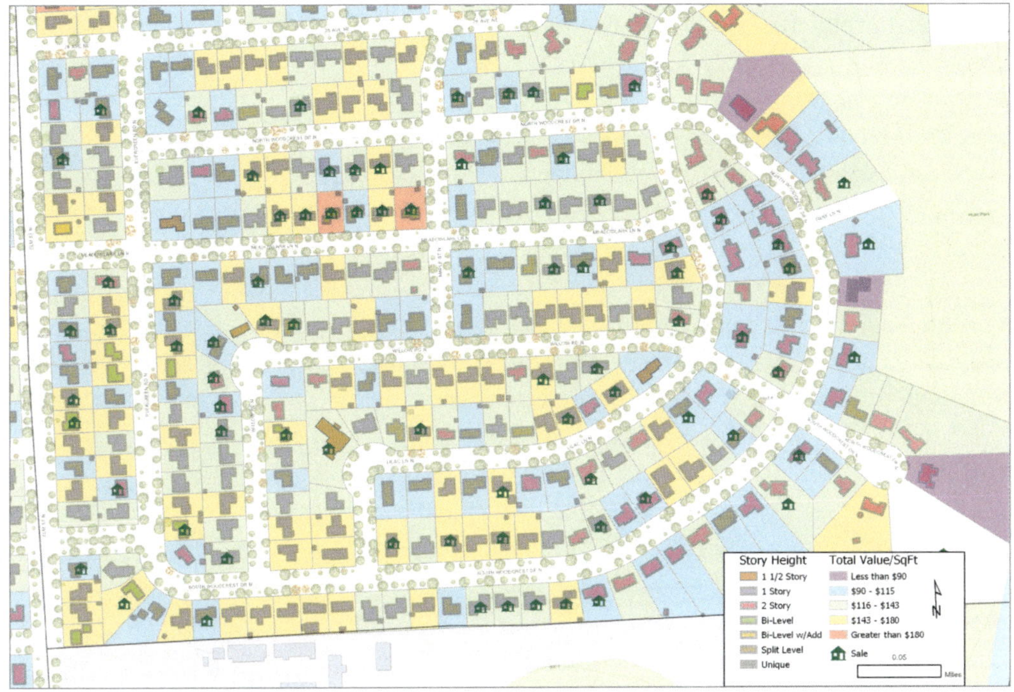

Figure 2.4: Static Map Example. *Source:* Esri and its licensors; City of Fargo, ND , Esri, HERE, Garmin, Intermap, increment P Corp, GEBCO, USGS, FAO, NPS, NCRAN, GeoBase, IGN, Kadaster NL, Ordnance Survey, Esri Japan, METI, Esri China (Hong Kong), swisstopo, OpenStreetMap contributors, and the GIS User Community.

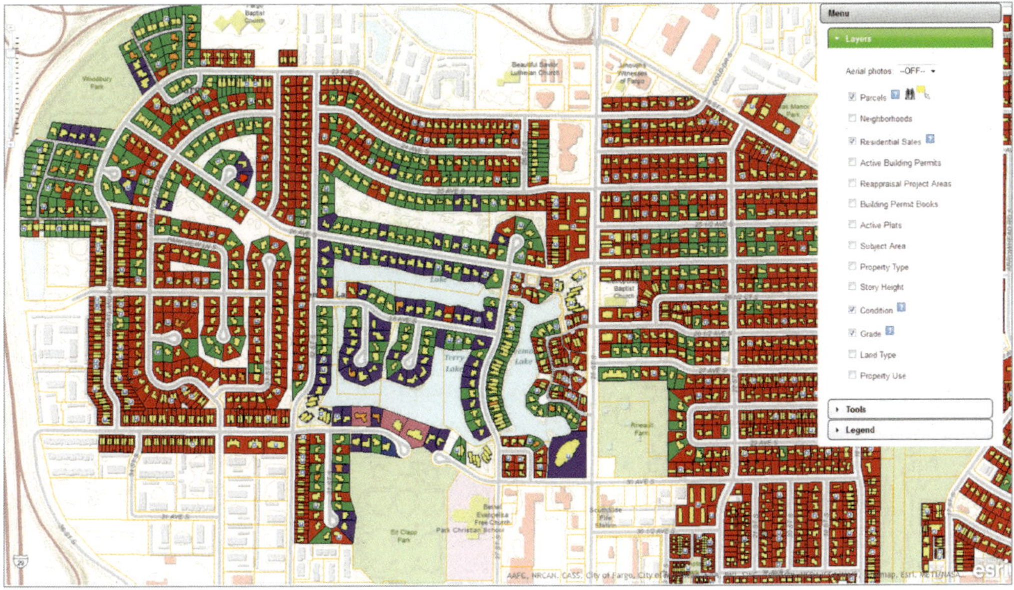

Figure 2.5: Example Web Map. *Source:* Esri and its licensors; City of Fargo, ND , Esri, HERE, Garmin, Intermap, increment P Corp, GEBCO, USGS, FAO, NPS, NCRAN, GeoBase, IGN, Kadaster NL, Ordnance Survey, Esri Japan, METI, Esri China (Hong Kong), swisstopo, OpenStreetMap contributors, and the GIS User Community.

- **Disadvantages:**
  - Scale-related problems arise when the user is attempting to visualize two or more properties that are relatively far away
  - Detail can be lost when attempting to plot a map of a large area, as parcel lines can become distorted or be too close together to properly visualize
  - Could require additional cartographic elements on the plotter version of the map in order to make them useful (e.g., a legend depicting map symbology, a north arrow, a scale bar, a title, additional text describing the data)

Dynamic maps are interactive, digital maps that can combine several visual layers, attribution, functions, and tools to aid in visualizing and analyzing an entire area. See *Figure 2.5* for an example of this kind of map. Characteristics of this medium are as follows:

- **Used for...**
  - A one-stop-shop for everyone's mapping needs (this is the preferred mapping medium for most assessor's offices)

- **Advantages:**
  - Can include various different layers of data
  - Contains the most up-to-date data available
  - Provides the most flexibility in use
  - More efficient than other mediums
  - Combines several web-based services online, and can be easily configured to display the data in any way the user would like to see it
  - Can be configured for use in the field on mobile devices, such as tablets or smart phones

- **Disadvantages:**
  - Most dynamic maps need access to the Internet to receive the most recent data
  - Users might require training in order to have the technical knowledge necessary to get the most out of the application

The use of web-based dynamic maps varies by jurisdiction, as some jurisdictions are more developed with respect to their GIS than others. However, the biggest benefit that a dynamic map brings to assessors is the ability to join spatial data with non-spatial data to display property data and other relevant information graphically, which then allows assessors to make better and more informed decisions regarding valuations. Dynamic maps also play a powerful role in data collection, as we will discussed in Chapter 3.

3.3 | Digital Mapping Program Best Practices

Several books and articles have been published on best practices for creating a digital mapping program, such as Roger Tomlinson's *Thinking About GIS: Geographic Information System Planning for Managers* (2013). Mapping programs in the assessor's office include several dynamic parts and are often integrated into a comprehensive local government GIS, where the GIS department administers the system and builds the products and services used by the assessor's office staff and the public. Several best practices for a mapping program within an assessor's office include the following:

- **Consider the strategic purpose and scope of the mapping program.** This refers to the primary use that the mapping program will have within the office—for example, to increase staff efficiency or to improve public engagement by having a more transparent government.

- **Plan and describe information products.** List out all potential solutions that will be used within the mapping program before it is implemented. These solutions might include one or several web mapping applications that serve a specific function within the office. They could also include a list of the types of data and data layers needed. The overall list should be detailed and should describe the purpose of the program, who will be responsible for maintaining it, and who its main users will be.

- **Data from outside sources (i.e., web services) should also be included in the list under what will need to be maintained.** The functionality of the program and how often data and services will be updated should also be listed. Potential assessor's office data products that might be considered are outlined in *Table 2.3*.

- **Create a data design and choose a logical data model.** The data model chosen could be an existing model or one that is defined in-house. A **data model** is a standardized method of maintaining data within a database that represents real world phenomena. Such a model could include property data to describe a parcel improvement and/or land attributes to describe the type of land on a parcel. Topological rules and relationships are also created to describe cadastral editing, as discussed in Section 4.3.

- **A good data model will allow room for expansion based on the need and demand for the phenomenon that is being described.** It should also provide a key identifier that relates to other tables within a database schema (e.g., a unique number, such as a par-

Table 2.3: Office Data Products

Parcel Lines	Parcel Centroids	Address Points
Taxing Districts	Road Networks	Plats/Surveys/PLSS
Building Footprints	Condominium Sites	Property Value
Tax Appeal Locations	Quality Grade Descriptors	Story Height
Building Area	Land Area	Demographic Data
Parcel Polygons	Orthophotography	Aerial Imagery
Other Property Data	Zoning Data	Soil and Slope
Sales Ratios	Equity Measures	Neighborhoods
Model Coefficients	Modeled Values	Assessed Values

cel or sale key). When maintaining parcel data, a data model provides rules for defining spatial relationships between data layers—these are known as "topological rules."

- **Implement the model and test the products.** This is where the rubber meets the road—the implementation of the plan and the creation/conversion of all data layers into the appropriate schemas. If web applications are being built, it might be necessary to test its usefulness and tweak the defined solutions where necessary.

Additional elements for the preparation and creation of a mapping program are outlined in Tomlinson's book. It usually takes an entire year (or two) before the effects of a new digital mapping program are fully understood. Benchmarks and performance metrics should be created to track the program's effectiveness from the very first day that it is adopted, and a management plan should be developed to ensure that all data are well-maintained within the database.

4 | Maintenance of Parcels in Cadastral Mapping

As the mapping program is implemented and a standardized parcel identification system is created, parcel boundaries and data must be updated on a regular basis. There are several methods for ensuring that this gets done, depending on how parcel geometry and data are maintained within the cadastral mapping framework. Having a well-documented and thorough maintenance program that takes into account ownership changes, cadastral layer changes, and general CAMA data and map layer maintenance is absolutely essential.

4.1 | Deeds and Deed Maintenance

A good cadastral maintenance program includes a method for analyzing ownership transfers though recorded documents such as deeds. A **deed** is a legal document that conveys conditions upon which property is transferred from one person to another. Deeds are usually filed and recorded at a County Recorder's Office and will often provide information on past and current ownership, a description of the land, and other conveyance information.

The type of deed used—and there are several different kinds—will depend on the type of ownership transfer that took place. However, it is important that a chain of ownership also be kept in a database that is updated as new deeds are recorded. Deeds will provide a description of the land being conveyed, which is an important comparison to the existing land description to ensure that ownership boundaries match up. Descriptions that don't match could require a boundary line adjustment. The *Standard on Digital Cadastral Maps and Parcel Identifiers* notes that procedures should be developed to ensure that this system is properly maintained.

4.2 | Parcel Boundary Transactions

Boundary lines are segments that connect to form a polygon, which denotes a land parcel unit. These lines are described using **legal descriptions** (a method for describing a fractional unit of land). This can be done in several different ways: via metes and bounds, lot and block, and/or rectangular descriptions.

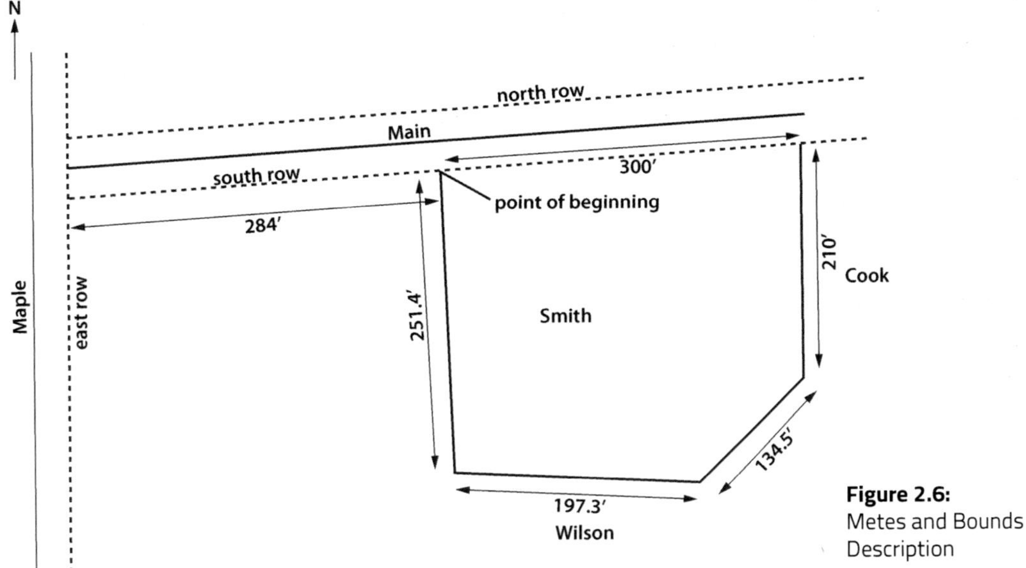

Figure 2.6: Metes and Bounds Description

Metes and bounds is a surveying method that identifies parcel boundaries relative to the distance from landmarks (Wade and Sommer, 2006). This was widely used by the original 13 states to describe irregular plots of land. Documented descriptions relied heavily on land features such as trees, rocks, streams, and other physical phenomena to describe the boundaries of a parcel. Modern uses of this system are still used—however, surveys always begin from well-known locations such as surveyed benchmarks, and precise bearings and distances are now the standard. A metes and bounds description contains several elements, which are described in the following paragraphs and represented visually in *Figure 2.6*. More detail is provided in IAAO's book, Property Assessment Valuation (2010).

- **Point of beginning (POB)** – This refers to a well-known location, such as a benchmark, to denote the starting point of a description.

- **Definite corners** – These are clearly defined points at a specified distance from known locations on the survey.

- **Length and bearing** – This refers to the length (in feet or meters) of a line that begins at the POB, and extends to a location at a bearing (or heading). The bearing is the horizontal direction of a point in relation to another point, expressed as an angle from a known direction such as north (Wade and Sommer, 2006).

- **Names of adjoining features** – Many descriptions provide distances from physical features to avoid as much ambiguity as possible.

- **Area** – The area of the described piece of land is given in square feet or acres.

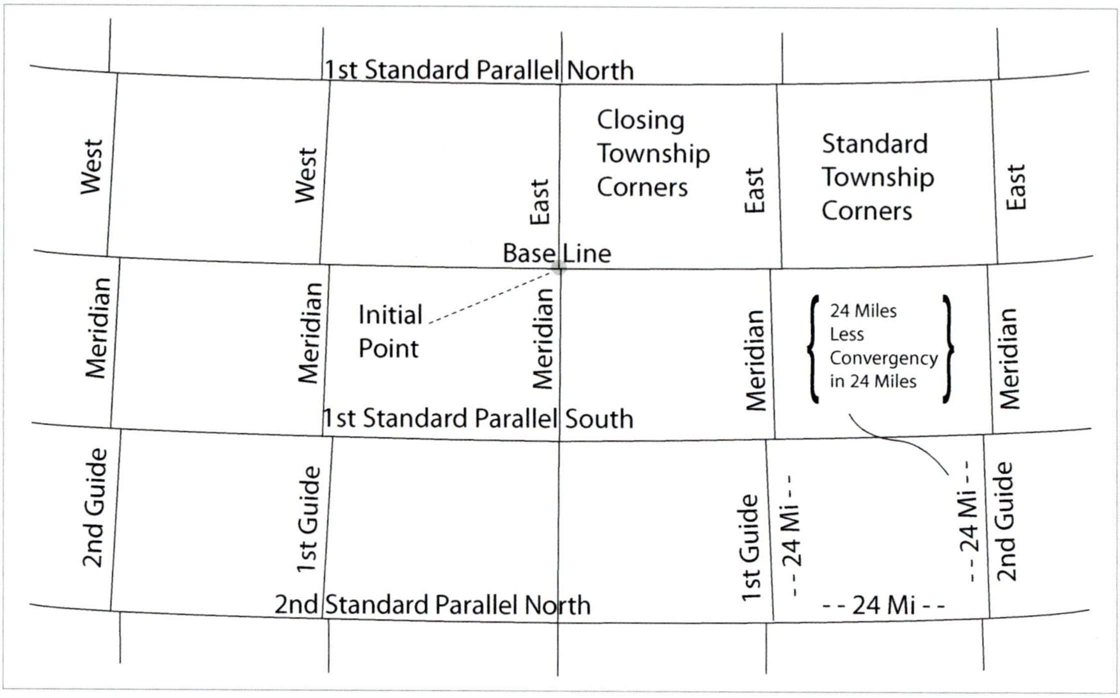

1st Standard Parallel North
Closing Township Corners
Standard Township Corners
West
West
East
East
East
Base Line
Initial Point
Meridian
Meridian
Meridian
Meridian
Meridian
24 Miles Less Convergency in 24 Miles
1st Standard Parallel South
2nd Guide
1st Guide
1st Guide
2nd Guide
- - 24 Mi - -
- - 24 Mi - -
2nd Standard Parallel North
- - 24 Mi - -

Figure 2.7: Base Lines and Guide Meridians

C A N A D A
M E X I C O
G U L F O F M E X I C O
ALASKA
UNITED STATES
DEPARTMENT OF THE INTERIOR
BUREAU OF LAND MANAGEMENT
PRINCIPAL MERIDIANS
AND
BASE LINES
GOVERNING
THE UNITED STATES
PUBLIC LAND SURVEYS
INTERIOR—GEOLOGICAL SURVEY, RESTON, VIRGINIA—1988

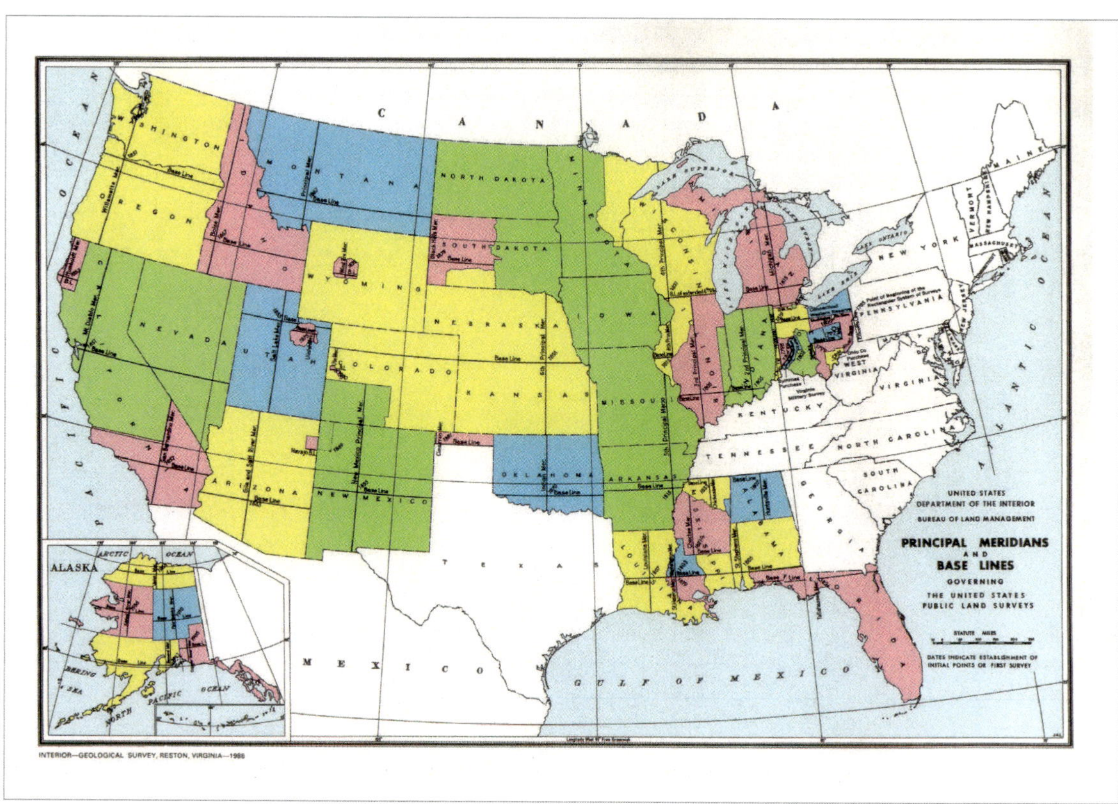

Figure 2.8: Location of Principal Meridians and Base Lines on a 1988 Map

The rectangular land survey system is often referred to as the **Public Land Survey System (PLSS)**, and was adopted by the federal government in 1785. The PLSS references principal meridians that run north to south and east to west, which make up an approximately 36-square-mile division of land called a "township." Each square mile of the township is called a "section" of land and can be broken into half- or quarter-sections, or subdivided even further into quarter-quarter-sections, and so on. See *Figure 2.7* for an illustration.

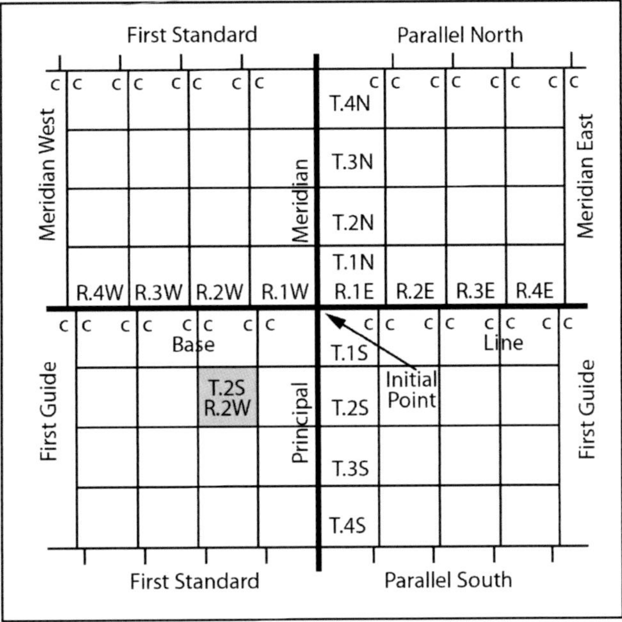

Figure 2.9: Range and Township Lines

Townships and sections are not always perfectly square since the meridians converge on each other. They are sometimes shifted at base lines and principal meridians (as shown in the *Figure 2.8*) in order to accommodate the curvature of the earth.

The place where the principal meridian and standard line intersect is called the **initial point**. Several counties throughout the United States have made it a priority to survey and find the exact locations of section corners, as they provide a great starting point as a control or benchmark for an accurate cadastral survey used to subdivide property.

Another surveying method that builds off of the metes and bounds system—the **lot and block method**—is the most common form of legal description, as it is the simplest to understand. This method references a subdivision plat, which is divided into rights-of-way, easements, and blocks and lots; and other important surveying elements. Blocks are numbered areas that are divided and separated by right-of-way or plat boundaries. Lots are subdivided blocks that are also numbered to form land parcels.

Subdivisions are often described using metes and bounds, and then referenced as newly created lots *within* blocks of a subdivided piece of land on the recorded subdivision plat document. The lots and blocks listed in the subdivision plat is often the new legal description. As an example, the subdivision in *Figure 2.10* is recorded as Cusack's subdivision, and the highlighted lot (Lot 7) is located in Block 2. Therefore, the legal description would be read as: Lot 7, Block 2 of Cusack's Addition.

Parcels are often described using at least one of the above methods, or through some combination of the two. The maintenance of parcel boundaries is essential in order for assessors to preserve an appropriate land unit that can be used to identify, list, and value properties for ad-valorem tax purposes. This maintenance system should include everything that is fixed to that land unit which might command value in the market.

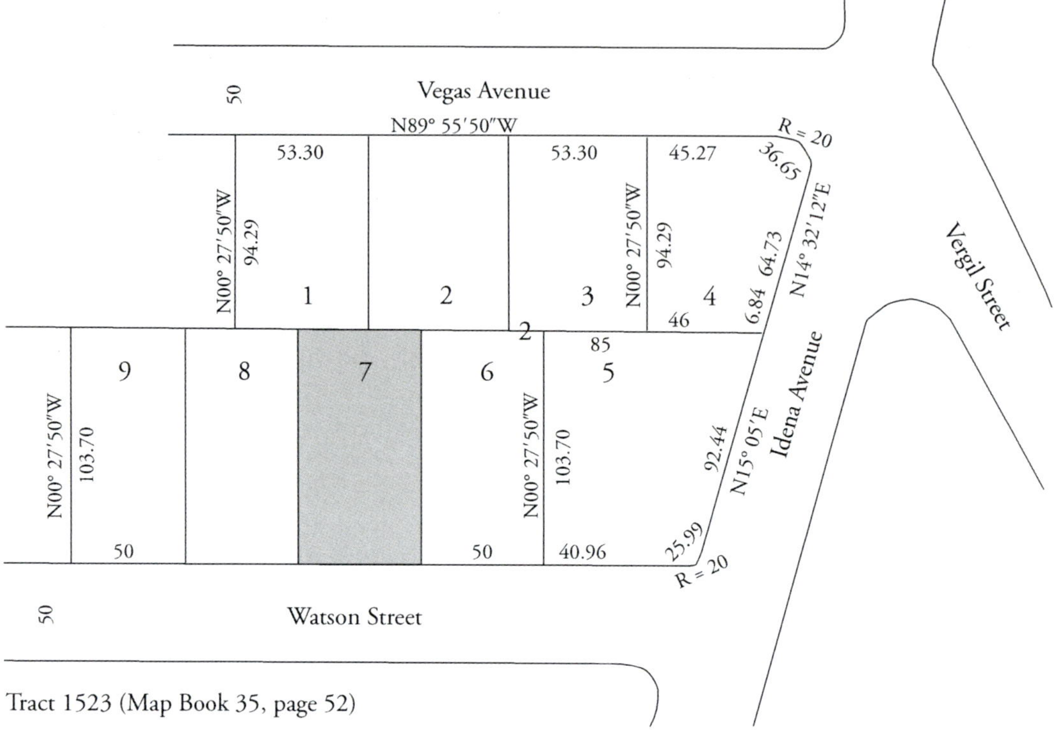

Figure 2.10: Lot and Block System

Maintaining parcel boundaries usually starts with the recorder's office when a deed or a plat is recorded with a legal description, and this responsibility eventually transfers to the assessor or GIS technician who draws the new boundaries within the county's base parcel framework.

4.3 | Drawing Parcel Geometry Using GIS

Prior to the use of GIS, parcel boundaries were drawn out on paper maps by professional surveyors. The evolution of modern technology has made editing parcel geometry and land records data much easier and more efficient. The creation of digital parcels is conducted through several mediums such as digitizing, coordinate geometry, topology, and parcel fabrics. Each method has benefits and drawbacks, but one or more of these methods are used by most jurisdictions throughout the United States. These methods are summarized as follows:

- **Digitizing** – This method takes manual data and uses a computer to convert them into a digital format. The user must manually draw the lines and/or points on the GIS to provide context for an approximate location. This could also be supplemented by geo-referencing and by layering scanned images or orthophotography. The drawback to this method is that it might not be very accurate and it is also very time-consuming.

- **Topology** - Topology provides a set of rules for how lines, points, and polygons share their geometry and connectivity. Some of these rules constrain how lines are created to maintain the integrity of the data. Some examples include not allowing overlapping lines, not allowing gaps between polygons, and not allowing for dangles or overshoots. Maintaining parcels using topology is an excellent way to ensure integrity of the database. However, it can be a time-consuming process to clean up any errors that do occur.

- **Coordinate Geometry (COGO)** - This method utilizes geometric descriptions such as bearings and distances. The bearings and distances often follow metes and bounds descriptions in order to provide an accurate plot of the parcel lines. The bearings and descriptions are often maintained with the line in the GIS database. COGO can be used in conjunction with topology.

  The drawback to COGO is that many metes and bounds descriptions are wrong or inaccurate and might not match with existing geometry. Thus, time-consuming research might need to be conducted, or an accurate survey completed, to ensure that the data are accurate. Otherwise, COGO does a great job of maintaining bearings and distances as attributes of the lines for future reference and symbolization.

- **Cadastral Fabric** - This method considers all of these methodologies. It consists of a network of connected parcels which are represented by point, line, and polygon features. These are all developed into a special aggregate network called the "fabric." The fabric can maintain various hierarchical levels of geometry within the same network, including tax parcels, ownership parcels, subdivision plats, survey control points, and PLS reference lines. Each aggregate level of geometry is built while keeping in mind the topological relationship within and between the various layers embedded within the fabric.

  The benefit of the fabric is that it can be adjusted with new and updated surveys. The fabric allows for the conversion of existing data and the ability to maintain a history of parcel geometry. It maintains COGO attributes within its database, along with a creation and obsoleting date.

Regardless of the method used to ensure geometry and relationships with lines, the main issue that many local governments are concerned with is the accuracy of their parcel layer. Some jurisdictions find and survey their PLSS section corners in order to use them as control points for drawing out parcels. Parcel fabric allows for hierarchal coding when locations (either known or less known) are set. Thus, most known locations will not move when adjusted. Parcel lines are enhanced through the annotation of their line distances and whether or not arcs are present. The angles and bearings are also of benefit to the public or to other government organizations.

After parcels lines are drawn, polygons are often created to form a fully enclosed space. This space, when attributed with the parcel identifier, can be joined with other databases to provide symbology for those associated records. This is of great benefit to the assessor in order to visualize property data, as shown in *Figure 2.11*.

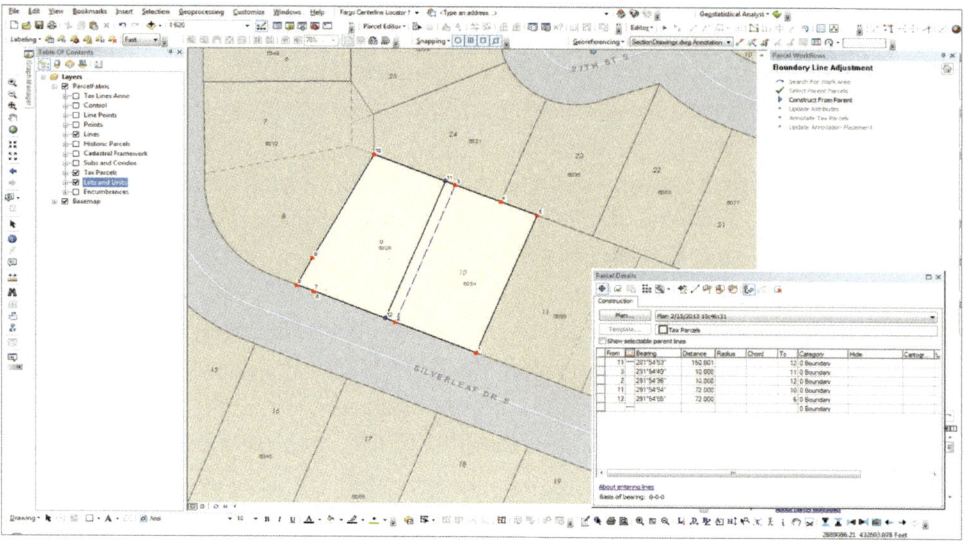

Figure 2.11: Editing Metes and Bounds Using Esri's Parcel Fabric. *Source:* Esri and its licensors; City of Fargo, Esri

Please see **Case Study 2.1** ("Parcel Fabric Mapping & Website Documentation: Tioga County") at the end of this chapter for a demonstration of how one county made use of this technology.

5 | Supplementing CAMA Data with Cadastral Mapping

Parcel geometry is significantly enriched by supplementing property data from a land records database, such as a CAMA system or land records database, to visually showcase data in a web mapping application. Often, land records are maintained in a database that is separate from a GIS database. In most cases, they're in another system entirely.

The problem that often arises when considering how to integrate additional data with cadastral geometry (parcel polygons) is that it can be tricky to update GIS databases from the other systems, since CAMA databases can be proprietary. In addition, the frequency of data updates often poses issues which depend on the amount of data that must be transferred from one database to another. Many jurisdictions set up a data extract or batch extract from CAMA that is then joined or updated within the GIS database on a monthly, weekly, or nightly basis.

The next few sections pose additional thoughts regarding the use and integration of land records, CAMA, and GIS.

5.1 | Land Records in Computer-Assisted Mass Appraisal Systems (CAMA)

CAMA systems are programs used by assessors to maintain land record data, to assist with the valuation of properties (using some or all of the approaches to value), and to administer all statutory requirements regarding the property valuation system for a jurisdiction. CAMA systems are essential because they help the assessor maintain the data necessary for valuation and listing properties efficiently.

CAMA systems often automate several processes, such as the calculation of values for mass appraisal based on a particular valuation approach. They also create tax rolls and run essential reports that are necessary for conducting the business of an assessor's office. Data are often changing quickly within CAMA systems to keep up with the land administration and valuation functions of the office. Valuation approaches most often included in a CAMA are cost, income, and market. Alternate approaches using regression are also often built within CAMA systems to enhance and reconcile with the other approaches.

Some CAMA systems have GIS applications built within them that can identify the locations of properties or perform basic mapping of general attributes. However, these systems often use a GIS database to pull data into the mapping application, rather than pulling directly from the CAMA system. This may be problematic since GIS databases are typically updated in monthly, weekly, or daily extracts. Separate mapping applications from CAMA that take an extract from a CAMA database and map it into a GIS database to perform similar and more advanced functions are also often maintained. The functions able to be performed will depend on the combined capabilities of the GIS and mapping program.

5.2 | Land Records in GIS Databases

Land records data in a GIS are essential to the identification and visualization of property data. The extraction of land records data from CAMA to a GIS database is generally the accepted methodology used throughout many jurisdictions. As mentioned, a data schema is mapped from the CAMA database to the GIS database, to which the data are either: (a) pushed from CAMA to GIS, or (b) simultaneously pushed and pulled, if edits are made from the GIS itself. Any data schema can be used, provided that it is flexible and meets the needs of the department.

A popular data schema among local governments is the local government information model created by Esri. A standard data model provides the same data structure for each jurisdiction that uses the tools and applications. The benefit to having a standardized data model is that mapping applications and their functionality can be easily shared. Efforts have been made to standardize parcel data schemas across local governments and up through the states, and at the federal level. We will talk about this further in Chapter 7.

5.3 | GIS and CAMA Integration

The idea of integrating GIS and CAMA has been around the assessment industry for the past decade. Prior sections in this chapter explained the issues that arise when trying to integrate data from CAMA to GIS (and vice versa), including, primarily, the timing of updates and the difficulty in pushing and pulling data from different data structures.

The benefit of having instantaneous updates is the ability to visualize or perform functions in a GIS without having to wait. Examples of this include mass editing in a web GIS and spatial editing. **Spatial edits** provide visual context to the data and ensure contiguity with data from other properties surrounding the subject property. Some mapping applications allow for the ability to select many parcels (e.g., a group of properties along a busy street to analyze potential impacts to value), and are also able to edit all parcels at the same time instead of having to go into CAMA, type in the parcel ID for each property, and

Spatial editing also provides geographical context that otherwise would not be found within CAMA systems, which are mainly records based. Thus, having a spatially enabled CAMA would benefit many assessment jurisdictions. Other benefits include running spatial data analysis (which might include calculations), and providing and displaying appropriate market comparables with recent sales data.

edit each record separately. This saves a considerable amount of time.

Although in most jurisdictions GIS and CAMA systems are able to talk with each other to some extent, some systems are more evolved than others. It will be beneficial in the future for CAMA and GIS vendors to work together to provide a best practice methodology for a more seamless and efficient integration.

6 | Conclusion

The administration of land is an important component of the assessor's workflow, as it involves describing the unit or boundary to identify, list, and value property. Components such as property records, parcel identification, deeded legal descriptions, valuation information, and other recorded information are pulled together to form a land administration system. This system, or workflow, is essential in local governments for making sure that cadastral data are being property maintained. The parcel geometry is used by several organizations and government entities for much-needed mapping and information products.

It is up to the assessor to ensure that parcels are listed on the tax rolls appropriately, and that each parcel is given a standardized identifier which distinguishes it from other parcels. Cadastral maps should contain accurate parcel lines which maintain some basic topology. Thus, jurisdictions must have a good digital mapping program which supplements data from multiple data sources with cadastral data.

Assessment offices are generally the authoritative agency for which most property information is obtained. It is therefore incredibly important to have a good GIS and CAMA data extraction methodology—one that efficiently moves data from one database to the other. Web mapping applications add several benefits to a well-integrated GIS and CAMA. Data accessibility and the integration with CAMA and other systems will be addressed in more depth in the next chapter, along with a discussion of how assessment professionals are creating methods to be more efficient for data collection and for the use of supplemental data.

Case Study 2.1

Parcel Fabric Mapping & Website Documentation: Tioga County, Pennsylvania

Author: Scott Zubek, GISP • Director, Tioga County GIS Department

PARCEL FABRIC DEFINED

The "fabric" or "Esri Fabric" is a connected system of GIS data types that form an optimal environment for accurate and streamlined parcel editing. Edits to all of these connected data layers occur as new information is entered into the system via the most accurate reference material available (e.g. recorded survey/deed defined with coordinate geometry, aka COGO). The fabric is loaded into a standardized data model for future expansion of its use with other datasets. This standardized data setup is referred to as the Local Government Information Model or LGIM. Along with providing the opportunity for standardization, the fabric, as loaded into the model, has the flexibility to be merged with local/customized data and overlaid with many other GIS layers.

ORIGINAL PARCEL MAP HISTORY

The Esri Parcel Fabric data model has replaced Tioga County's now retired GIS parcel map layer, which was created circa 2003 using standard methods of the time. The original parcel map was created by digitizing tax plate mylars into the GIS environment, which is a common practice. In the time since the creation of the original GIS parcel map, the evolution of the technology and methods for cadastral mapping in contemporary GIS has drastically changed, thus making the creation of a thoroughly researched and highly accurate fabric-based parcel map possible. In the fall of 2016 the GIS and tax assessment departments went fully "live" with the fabric-based data, which comprised a complete redraw and deployment of the entire parcel base-mapper the standards outlined above.

It should be emphasized here that even with the increased accuracy of the data per the new fabric-based mapping system, the data maintained by the county and other agencies, including web-based data, is designed to serve as a *secondary* representation of real property found within this jurisdiction as compiled from the most accurate assessment, emergency services, and other public records available. Users of the data and any other related reference information are hereby notified that these primary sources should be consulted for verification of the information presented. Tioga County, its staff, and the staff of the various agencies who develop & maintain the data & software assume no legal responsibility for the information referred to in this section or in the rest of the document.

COUNTY GIS INTEGRATION

- **Parcel Conversion and Deployment –** The parcel conversion process constituted a county-wide redraw of the cadastral lines by a professional vendor that used a standardized system of ranking parcels by accuracy to reflect the best recorded reference information/geodetic control available; whether it be a parcel's most recent recorded deed or a recorded field survey with ground coordinates. The rest of the parcel map was then built by anchoring the parcel build process to those higher accuracy areas based on the recorded documents. The initial redraw was completed in early 2016 and then continually maintained in the fabric environment by Tioga County staff.

- **CAMA Data & Automated Updates -** The parcel base map layer that will be used for county-wide mapping and web services is automatically updated with information exported from the computer-assisted mass appraisal (CAMA) database system on a daily basis. This GIS layer will also be available as a web services in Tioga County's rerelease of its GIS base map viewer.

- **Right-of-Way (ROW) Mapping -** The parcel fabric environment provides a different way of looking at the data. The parcels have been redrawn with the information contained in a deed and/or survey. Therefore, in some cases, parcels are defined to the center of a right-of-way (ROW) and created in the GIS environment as such. A set of "cadastral" layers will be available, which will define the location of Right-of-way lines with respect to parcels.

  Tioga County has a unique dataset in this regard as many of the ROWs in the new map have been drawn using PENNDOT provided ROW maps that County staff was able to retrieve from PENNDOT'S archives. This will also be apparent on the website as ROW lines will be seen as "shading" over the parcels. This will greatly enhance visual accuracy of the data and also can also be utilized for practical purposes, such as setback calculations.

DATA EXPORTS

The export of fabric data for exchange, or other uses, will be no different in practice than previous GIS exports. However, additional layers will accompany the exports that will define the location of ROWs per the ROW mapping that was completed as noted above. The export format will typically be in a file-geodatabase format, but all standard GIS export and exchange formats (e.g., shapefile) can be accommodated. The GIS department only includes parcel/ROW lines and very basic ownership information with its export. A version of the export from the GIS software, which includes detailed CAMA data, is available upon request.

REFERENCE PARCEL DATA INCLUDED WITH STANDARD EXPORT

The GIS parcel fabric standard data export includes two additional layers for reference, which are included for the benefit of the requestor. The two additional layers have been geo-processed via two versions of the county's rights-of-way, based on a clip of the output parcel polygons. One version is a clip based upon the researched cadastral rights-of-way that was completed with the fabric redraw process. The other version of the fabric has the EMS-911 right-of-way areas clipped from the parcel polygons. The EMS-911 right-of-way clip layer will match the road centerlines that the county uses for dispatch & routing. The cadastral clip layer is based on property ownership research and will match the cadastral right-of-way centerlines. The county's standard export format is a file-geodatabase. The reference layers are contained in feature datasets within the geodatabase, along with other reference layers, with the intent of creating an orderly schema for the end user.

WEB SERVICES

The GIS/CAMA layer will be available for public viewing via an ArcGIS Online formatted web map/app, accessible on Tioga County's government website. The web service will be automatically updated with both parcel edits and CAMA data on a daily basis. The service will also display many other County GIS layers and allow the user to perform simple queries on the data to find a particular parcel or area of the map for further viewing. The new fabric-based web viewer was released in November of 2016.

GIS ATTRIBUTE TABLE – FIELD SCHEMA NAMES & DESCRIPTIONS/ALIASES

The table below is an index of the GIS attribute default table field names (as part of the schema), descriptions, and the corresponding aliases that the county typically uses for web services and general mapping purposes.

- Schema field names in bold/italic indicate a field included in the standard GIS export. All other fields are considered part of the CAMA detailed export as noted above and are not part of the standard GIS export.

- Field names greyed-out indicates a field that is not included in GIS export or the detailed CAMA/GIS export. Staff has decided to leave these fields in the list in the event that we make that information available at some point in the future.

- The Tioga County GIS department and assessment department reserve the right to redact and/or add fields to this list and corresponding export at any time as we continue to work in the new fabric environment going forward. We also reserve the right to modify the "Field Alias/Description" values, as this field is dynamic per the needs of our customers. It should be noted that, for the most part, the schema design and descriptions/aliases remain unchanged.

Table 2.4: Index of GIS Attribute Default Table Field Names

SCHEME FIELD NAME (DEFAULT)	FIELD ALIAS/DESCRIPTION	DATA TYPE (DEFAULT)
OBJECTID	Geodatabase Default Field	Object ID
PARCELID	Parcel Identification Number (PIN)	TEXT
MBLU	PIN (CAMA Format)	TEXT
REM-MNC	N/A	NUMERIC-LONG
REM_PID	Vision PID	NUMERIC-LONG
REM_PIN	Vision PIN	TEXT (35)
REM_ACCT_NUM	***Vision Control Number***	***TEXT (20)***
REM_GIS_ID	***PIN (GIS Format)***	***TEXT (30)***
REM_ALT_PRCL_ID	GIS ID (Retired Format)	TEXT (35)
REM_MBLU_MAP	District Number	TEXT (7)
REM_MBLU_MAP_CUT	N/A	TEXT (3)
REM_MBLU_MAP_BLOCK	Plate	TEXT (7)
REM_MBLU_MAP_BLOCK_CUT	N/A	TEXT (3)
REM_MBLU_MAP_LOT	Parcel	TEXT (7)
REM_MBLU_MAP_LOT_CUT	N/A	TEXT (3)

CHAPTER 3
Data Collection, Verification, and Measurement

1 | Introduction

Data are essential to listing, identifying, and valuing property. Assessors are often the primary source of truth for many local government agencies and public entities with respect to the use and dissemination of property data. Data are the life blood of local government, and even more so in the process of administering a fair and equitable **ad valorem** tax. "Ad valorem" refers to a tax according to value, which is the basis for how property taxes are distributed within the United States.

Sale transaction data are especially important to understanding the dynamics that shape the local markets. Maintaining a separate database that includes a data snapshot of a property at the time the property sold can provide assessors with valuable information pertaining to the cost or demand for a particular amenity compared to other properties that have sold. There are numerous methods that can be utilized in maintaining data in database structures, but these are not the focus of this book. Instead, we will discuss the ability to collect, compare, and map these characteristics.

Assessors may collect data from a wide array of sources, including physical inspection (accomplished by walking or driving through a neighborhood), imagery, and multiple listing services. The important aspect of data collection is ensuring that the data obtained for CAMA records are accurate. Accurate data are important to the valuation of properties because they are the primary means for comparing one property to another. Cadastral data, as discussed in Chapter 2, are important to finding the location and boundaries of the property. In this chapter we will discuss how property data, and other associated data, can assist in defining the quality and consistency of a property when deriving valuations.

The essence of data collection, verification, and measurement is the focus of this chapter. We will discuss best practices and common methodologies utilized across the profession with respect to data types within the context of mapping, including how these data are collected and how they can be used effectively. In addition, we will cover some best practices for mapping both discrete and continuous data, and the visual analysis of these data.

2 | Data Maintained by Assessors

Data collection is one of the most expensive and most important tasks conducted by an assessor's office. The several types of data that assessors collect are gathered through either primary or secondary means. Primary data collection includes those types of data that are collected solely by the assessor (e.g., property data that have been measured or observed directly). Secondary data are data obtained through other sources, such as a multiple listing service or website with property listings. This could include economic data or census data collected through the Internet.

IAAO (2010) breaks out data that are to be "collected, analyzed, and processed" into three overarching categories which are also useful within the context of GIS: (1) general data, (2) specific data, and (3) comparative data. A fourth group of data includes imagery and remotely sensed data which provide assessors with visual and graphical representations of the property at a given point in time. Each of these groups of data can contain spatial, non-spatial, imagery-based, and/or remotely sensed data, along with other representations of data, such as 3-D data. *Table 3.1* provides an overview of all of these different types of data.

Table 3.1: Types of Data Utilized by Assessors

GENERAL	SPECIFIC	COMPARATIVE	IMAGERY	REPRESENTATIONS
Physical (Environmental)	Site	Cost	Aerial	LiDAR
Economic	Off-Site	Sales	Oblique Aerial	Vector based data
Governmental	Improvements	Income/Expense	Street Level	3-D Models
Social			Multispectral	

2.1 | General Data

General data include trends or attributes that affect property value on the national, regional, and neighborhood levels. These types of data are often important to describing general market forces or in aggregating various descriptive data. General data include physical, social, environmental, and governmental data; neighborhood data; and other trends.

- **Physical** or **environmental data** are often composed of spatial boundaries that delineate the locations of significant developments from natural phenomena, such as rivers or lakes, and man-made phenomena, such as major roadways. Physical location data are often shown on maps and are perhaps the most important value-influencing attributes. Other physical data types might include climate, topography, accessibility, and so on.

- **Economic data** are applied to valuation-related studies to help understand the supply and demand forces inherent within the local market. Economic data often provide information on a resident's ability to afford a property, and therefore require looking at indicators such as mortgage and lending rates, employment, occupancy rates, rental rates, market costs, capitalization rates, and so on. Economic data are often obtained through government agencies.

- **Governmental data** often include information on how statutes or regulations imposed by local governments (or other levels of government) might influence property values. Government factors might include laws that restrict what can be built where, what types of occupants can live on a property, or the number of parking stalls allowed on a property.

  Zoning often restricts the type of properties that might be built in a certain location in order to prevent urban sprawl or other development issues that might cost taxpayers in the long run. Access to emergency services also affects property values, especially with respect to fire protection. Governmental data might also

Data that do not have an absolute location associated with them are called *non-spatial data.* Non-spatial data are typically collected by assessors though the visual inspection of a property and then identified through an address or Parcel Identifier, which could then be joined with spatial data for display on a map.

include political boundaries, or assessment neighborhoods delineated and collected by government officials.

- **Spatial data**, often referred to as **geospatial data**, are those types of data that provide an absolute location for an earth-bound phenomenon. In valuation, this often refers to the parcel or sale, and is often stored in a database as an XY coordinate within a spatial reference system or as topology, as described in Chapter 2.

- **Social data** are also closely related to the economy. These data help gauge a population's demand to cluster in certain neighborhoods or economic aggregations. This is very much a spatial phenomenon that considers the opoulation factors of an area, including population density, education, crime and more. Many of these variables can be found through a government agenciesand may be aggregated by submarkets (e.g. block, block group, ward, postal code). Social data can often be beneficial to assessment professionals. When joined with assessment data, it can help serve as an additional step of quality control to ensure valuations are equitable across all subgroups.

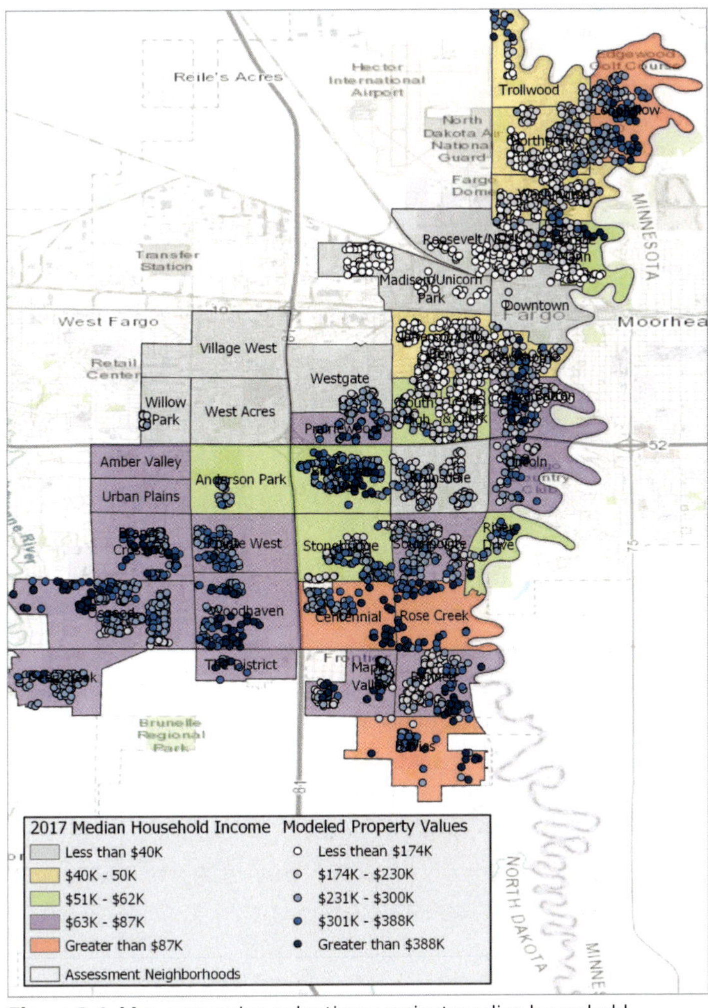

Figure 3.1: Map comparing valuations against median household income. *Source:* Esri and its licensors; City of Fargo, ND , Esri, HERE, Garmin, Intermap, increment P Corp, GEBCO, USGS, FAO, NPS, NCRAN, GeoBase, IGN, Kadaster NL, Ordnance Survey, Esri Japan, METI, Esri China (Hong Kong), swisstopo, OpenStreetMap contributors, and the GIS User Community.

IAAO (2010) provides a detailed list of examples of data that fall into the categories of physical, economic, governmental, and social (PEGS). These are broken out in *Table 3.2* for easy reference, and the examples encompass residential, commercial, industrial, and rural property.

Table 3.2: Types of General Data Used by Assessors

PHYSICAL	ECONOMIC	GOVERNMENTAL	SOCIAL
Appearance	Jobs	Municipal services	Population
Size, shape and area of lots	Family income	Planning	Crime
Street patterns	Price and rent levels	Zoning	Education
Proximity to support facilities	Ownership/tenancy ratios	Building codes	Neighborhood cohesiveness
Topography	Rents	Development regulations	Availability of labor
Climate	Turnover/vacancy	Taxes	
Utilities	Land use patterns	Special assessments	
Soil conditions	Amount of vacant land	Other policies	
Irrigation	New construction		
Hazards, externalities	Lender policies		
Traffic patterns	Foreclosure rate		
Waste disposal	Assessments		
Transportation access	Property taxes		
	Utility costs		
	Insurance rates		
	Direction of growth		
	Business turnover		

The application of these "PEGS" data, as they are normally called, provides context to the location and quality of the neighborhood being valued. Good quality neighborhoods generally have lower crime rates and higher levels of education, are near parks and trails, and have fewer rental properties. They can be bounded by rivers or lakes or buffered between high-end commercial developments.

The use and knowledge of PEGS data are essential to developing an opinion of value, and are also useful for understanding the effects of the valuation on these populations. Many jurisdictions maintain access to these data through the aforementioned sources or are able to visualize them on GIS. Many sources provide these data through **web services** where jurisdictions can connect to a website and pull the data directly into an application. A web service is an XML-based information exchange protocol that use websites to transfer data from one site or application to another. This is a common form of information sharing that allows jurisdictions to not have to store the data on their own servers.

▸ 2.2 | Specific Data

Specific data often include site, off-site, and improvement data on a subject's land and dwellings. This type of data is often the most important, and is usually the most accurate. Specific data on property attributes are typically collected by the assessor in the field or through other means, such as imagery. Special attention is given to the overall condition and functional utility of this type of data, as it is often utilized in mass appraisal models and used as units of comparison in the sales comparison approach.

Specific data can be further broken down into three different types: site-specific, off-site-specific, and improvement. **Site-specific data** include the classification, condition, and configuration of the property. **Offsite-specific data** include externalities that may directly affect the subject, such as whether it has access to city water, has a paved or gravel street, and so on. **Improvement data** include data on the property itself that relate to value. These could be construction grade, condition, year built, type of siding, size of the dwelling, and number of floors.

Table 3.3 contains a breakdown of common land characteristics that are collected by assessors that could have an effect on its valuation. The following table, *Table 3.4*, lists types of improvement data that are often used by assessors.

Table 3.3: Types of Land Data Collected by Assessors

LAND ATTRIBUTE	DESCRIPTION
Area	The size of the land parcel generally (square feet, acres, hectares, etc.)
Land-to-building ratio	The ratio of the footprint of the dwelling to that of the land area
Land use	The primary land use of the parcel
Topography	The amount of terrain or slope of the parcel
Soil	The type of soil on the parcel
Front footage	The length of front footage to the street
Water indicator	Indicator to denote that the property backs on to a body of water
Water frontage	The length of water footage
Irregular indicator	Indicator to denote that the property shape is irregular
Others	May include view or access properties, including easements or rights-of-way

Table 3.4: Types of Property Data Collected by Assessors

PROPERTY ATTRIBUTE	DESCRIPTION
Quality grade	Quality of materials (workmanship) utilized in construction
Size	Floor area measured in square feet; sometimes separated by components of the property
Year built	Year of construction
Condition	State of repair of the property or component
Story height	Number of floors
Siding	Type of siding of the property (e.g., vinyl, metal, brick)
Roof covering	Type of roof covering material
Type of basement	If basement is a walkout basement or a full basement
Frame type	Type of framing
HVAC	Type of heating or air conditioning system
Foundation	Type of foundation
Number of baths	Including half and full bathrooms
Number of fireplaces	Sometimes include type (e.g., gas, wood, stove)
Deck area	Area in square feet of deck or porch, or if large or small
Garage area stalls	Area in square feet of garage or number of stalls
X and Y data	Location data of the property in the form of latitude or longitude or UTM coordinates
Address	The situs address of the property
Parcel identifier	The unique parcel or tax identifier
Wall height	The height of a wall of a commercial property
Others	May include swimming pools, outbuildings, other value-influencing data

The **quality** of data is important, as it relates to how consistently the data are being accurately measured and recorded. It is essential that all assessment office staff have relevant training in data collection to ensure uniform data collection practices. It is also essential that properties are coded with appropriate codes relevant to the jurisdiction policy. Additional data collection practices can assist with the quality of data, including imagery. We will discuss this further in the next section.

The **quantity** of data is vital to ensure that data are collected in an efficient, yet accurate, manner. This is usually characterized by assessors planning revaluation areas in advance of data collection. It's essential that all of the data that are available on value-influencing characteristics be gathered and evaluated. Collecting insignificant data could place time constraints on assessment staff, which will certainly be felt at the end of the cycle. Performance measures and metrics on data collection methodologies should be monitored, and the data should be audited to ensure that staff are as productive as possible.

The ability to associate property data with spatial data provides additional context for the assessor to compare with other property data in the neighborhood. In order to make data spatial, the data must either be: (a) joined with data that are already associated with geographic coordinates based on a common field, or (b) have spatial coordinates already in place in the database. Many jurisdictions display property characteristics on a map so that staff can analyze the quality, consistency, and quantity of the data collected.

2.3 | Comparative Data

Comparative data often include data on sales, cost, or income. Sales are indicators of how a market is performing; thus, the sales data are essential for generalizing the valuation of a sample of properties on the population. Unsold properties are often compared against sold properties in order to develop an analysis of the differences in data characteristics, which can then be used to make an estimate of their contribution to value. This is arguably the most important type of data that the assessor collects, and it is essential that it be kept separate from actual CAMA data.

Sales data include attributes such as those listed in *Table 3.5* and the CAMA attributes listed in *Tables 3.3* and *3.4* as of the time of sale. The CAMA data associated with the sale are normally maintained in a separate database as they are a snapshot of that property at the time of sale. This will be used to provide additional context to the effects of the property characteristics on the sale price. The transactional sales data are often used to determine whether the sale is an **arms-length transaction**—that is, one that has been exposed to the open market and meets the definition of market value.

Table 3.5: Types of Sales Data for Residential Property Sales

SALE ATTRIBUTE	DESCRIPTION
Sale price	The sale price of the property
Adjustments	Amount and description of any adjustments to the sale value, including paid points
Sale date	The date that the property was sold
Instrument type	The type of instrument (deed) used to transfer the property
Number of parcels	Number of parcels involved in the sale
Sale type	Whether the sale involved improved or vacant land
Others	

Cost data typically include information on the cost of construction materials, in order to develop rates for the cost approach. Many jurisdictions subscribe to services that provide manuals with this information, or develop them in-house by using local construction companies or deriving them directly from the market.

Rental and expense data are essential to valuing properties using the income approach, and are typically gathered for income-producing properties. Income data, including rents, can be obtained through sales (deeds), leases, offers or listings, property managers or tenants, real estate brokers, private data organizations, or published data sources. If necessary, it can also be obtained through court appraisal exchanges, websites,

Figure 3.2:
Map of sales locations displaying the assessed value to sale price ratio.
Source: Esri and its licensors; City of Fargo, ND , Esri, HERE, Garmin, Intermap, increment P Corp, GEBCO, USGS, FAO, NPS, NCRAN, GeoBase, IGN, Kadaster NL, Ordnance Survey, Esri Japan, METI, Esri China (Hong Kong), swisstopo, OpenStreetMap contributors, and the GIS User Community.

or other jurisdictions, though there could be some privacy implications that the assessor will need to be aware of.

The spatial applicability of comparative data is very important to assessors as it provides, first and foremost, the spatial distribution of sales locations in a study area. Effectively, this shows how active a market is with respect to property transactions by illustrating how dense a location (such as a neighborhood) is with those transactions.

There are many different ways to display sales data on a map for visualization purposes. The sale record needs a location identifier, such as spatial coordinates or else through a common joined attribute. The two most common methods are summarized below:

- **Global Positioning System (GPS) coordinates**- Often referred to as **XY coordinates**, these provide the exact location (e.g. latitude and longitude) on the earth where a phenomenon occurs. These coordinates can be used in any reference system, including, often, point data or sale locations. Many CAMA systems are able to store this information within their tables and then plot it on a map.

- **Geometry (polygons)**. Sales can often be joined with spatially referenced geometry in order to display sale characteristics. Parcel geometry, as discussed in Chapter 2, is typically maintained through cadastral mapping. Data are typically joined to the geometry though a common identifier to display data attributes. Data can also be aggregated through various functions (e.g., sum, divide, average) at other levels of spatial hierarchy or scale, such as a neighborhood polygon, in order to provide generalization.

Figure 3.3 shows an example of a map that depicts aggregated data for a specific neighborhood.

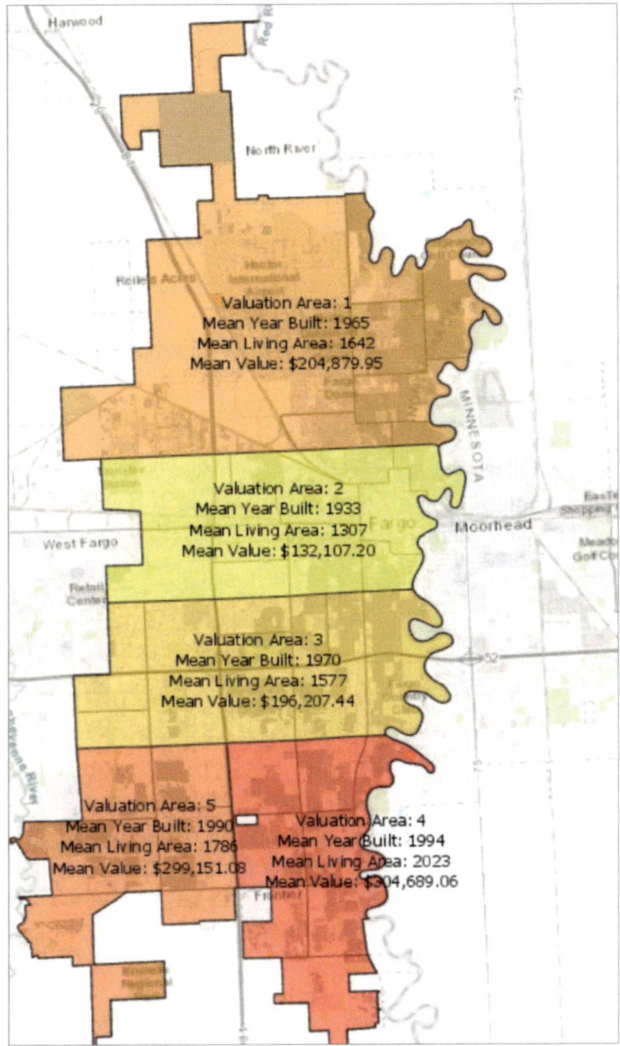

Figure 3.3: Map of aggregated data by valuation district. *Source:* Esri and its licensors; City of Fargo, ND, Esri, HERE, Garmin, Intermap, increment P Corp, GEBCO, USGS, FAO, NPS, NCRAN, GeoBase, IGN, Kadaster NL, Ordnance Survey, Esri Japan, METI, Esri China (Hong Kong), swisstopo, OpenStreetMap contributors, and the GIS User Community.

2.4 | Imagery and Remotely Sensed Data

Other data that assessors collect, but that are often obtained from outside sources or performed as services by vendors, include imagery and remotely sensed data. This might include aerial imagery, street-level imagery, oblique aerial imagery, or other kinds of multispectral imagery. Imagery is essential for assessors to quickly develop an opinion of the quality of a property's construction, and to compare subject properties with others that have already sold.

Imagery has provided an alternative means for data collection for assessors and is being used more and more widely by jurisdictions every year. The IAAO *Standard on Mass Appraisal of Real Property* (Section 3.3.) provides a rationale for the use of imagery in place of physical inspection, provided that the images used are no more than two years old in rapid growth areas, and no more than six to ten years old (as of the date of appraisal) in areas with slower growth.

The IAAO standard currently suggests that the images should have a pixel resolution of 6 to 12 inches (15.24 to 30.48 centimeters). Several jurisdictions have taken advantage of this standard and have developed custom applications in order to make their data collection processes more efficient. Section 3 focuses on the different types of applications and integrations that can be developed.

In this section, we will provide definitions for the various types of imagery and remotely sensed data that are often used by assessors.

- **Aerial imagery** is imagery that is taken from a platform on an aircraft. These images are often used as a cartographic data base map for interpreting urban and environmental features on the ground. Assessors can use these data to analyze land use and visualize the occupancies of the land, buildings, and surrounding areas.

When a multispectral camera is used to take imagery, the imagery can be manipulated for other urban and environmental purposes, depending on the spectral frequency across the electromagnetic spectrum. This type of imagery is often useful for agricultural productivity or for identifying land use patterns. **Multispectral imagery** also has various other uses in analyzing urban land cover and changes in the health of crops.

- **Oblique aerial imagery** is imagery that is taken at an angle perpendicular to the ground (usually 45 degrees). These images are often flown and taken at the same time as the aerial imagery from the same platform. Oblique imagery is important to assessors because it captures each side of the property from an angle which also allows them to capture other neighborhood characteristics. Higher resolution oblique imagery captures the quality of construction, type of property, and any other buildings or items on the property that may contribute to the property's value. Oblique imagery provides assessors with the opportunity to explore and better understand the environment under examination. Additionally, it lends context to the economic conditions

Figure 3.4: Aerial and oblique aerial images. *Imagery:* © 2018 Google; *Map data:* © 2018 Google, United States. *Image capture:* Sept 2013, 2018 Google, United States

that might underlie these neighborhoods and the factors that potentially contribute to those conditions. Other uses include examining vacant lots, visualizing commercial development configurations, and verifying building and construction types. Figure 3.4 shows an example of an aerial photograph and an oblique aerial photograph, in order to highlight the differences between the two.

- **Street-level imagery** is a great supplement to aerial and oblique area imagery as it provides a high-resolution street view of the property. Many applications provide

Figure 3.5: Street level image application

street-level imagery in a seamless, interactive display that allows the user to advance forward or backward down the street. It also lets the user zoom in on properties to better view the condition of exterior attributes and use measurement tools. Other tools—such as those that give the ability to extract images and add them to reports, or ones that let the user sketch within the application—are also growing in popularity. See *Figure 3.5* for a sample image taken at street level.

Imagery is often most helpful when used in conjunction with other layers—such as parcels, road networks, or even the location of sales—to provide a depiction of the neighborhood under examination. Being able to examine assessment data with imagery allows assessors to visualize and compare property for valuation purposes. As the resolution of imagery improves, so does its usefulness within the assessment industry.

2.5 | 3-D Representations

Although 3-D representations are not necessarily a data type, their potential importance in mass valuation of condominiums and skyscrapers is notable. These representations are created through the use of remotely sensed data, such as LiDAR, and created in GIS. 3-D representations can be used for several types of analysis, such as line of sight, to derive a valuation factor based on the adequacy of the view of the property owned by an individual. Appraisal theory suggests that the better the views, the higher a prop-

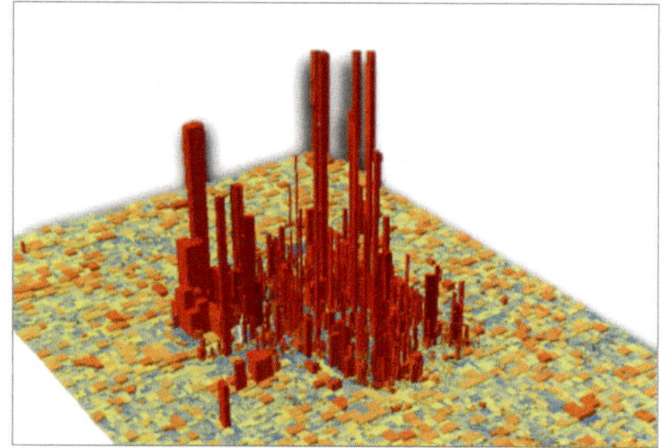

Figure 3.6: Example of 3D Value Map. *Source:* Esri and its licensors; Brent Jones, PE; Discovering "Insightful" patterns within your data using Business Intelligence Tools and GIS; IAAO's International Research Symposium & Forum, April 4-6, 2018, Boscolo Prague Hotel - Prague, Czech Republic

erty's value. 3-D representations can also provide a way to identify air rights or parceling within a high-rise property, and can be used for displaying those characteristics or divided spaces much like in a two-dimensional parcel map. Visualization in 3-D can also display the spatial distribution of values to analyze the variation from property to property if the assessed value is used as a measure of height (z-value) in rendering. 3-D rendering has come a long way over the past several years due to the intense computer processing that it employs. Its use in property assessment is gradually becoming more useful in the context of larger cities that have the resources necessary to maintain a 3-D cadastre.

3 | Data Collection Methodologies

The primary methods used by assessors to capture data have evolved considerably since their inception. Data has gone from being stored on paper card-stock forms to computerized relational databases, to being stored in enterprise spatial database structures. The evolution of the CAMA system has brought about several methods for collecting data that enhance the efficiency of staff, both in the field and at the office. The following sections describe several of these methods and offer some best practices in building a data collection program that saves time and tax dollars. It should be noted that many states have statutory obligations that need to be met and might not coincide with some of the methodologies discussed in this section.

At one point in time, **property record cards** were the primary means of maintaining cadastral records on a parcel. These cards, usually printed in card stock, provided a way for assessors to organize parcel information geographically or to perform stratification by property type or taxpayers.

The cards themselves contained legal descriptions, property information, and assessment information. Photos were often attached to the card, in addition to a sketch inscribed inside. The layout of the cards depended primarily on the importance of the information presented and the amount of space available.

The disadvantage of property record cards was that they were the only source of information for parcels and were distributed to appraisers who might misplace them. Thus, if was often difficult to get access to these cards, and maintaining them to ensure up-to-date information was also quite a chore.

3.1 | Manual Data Collection

Manual data collection is the process of collecting property data through non-digital means. Data collection consists of writing property data onto a paper worksheet, and then digitizing it into a computer system for later consumption. Resources needed for this collection method include a pencil, clipboard, well-maintained vehicle, tape measure, paper maps, and property cards. Prior to digital databases, assessment offices collected data on a paper card stock folder called a **property record card**. These cards were often kept in office filing cabinets where staff would be allowed to check them out to take into the field or perform other property maintenance duties.

This methodology poses several logistical and quality assurance issues, such as potential data loss due to the misinterpretation of written codes or notes. Additionally, the digitization process introduces another opportunity for error through misinterpretation or the accidental dropping of material as records are transferred. Often, property record cards are scanned into document management systems to ensure that data are not completely lost. Data errors might also occur with measurements taken at the property, as issues with consistency might arise depending on how the appraiser measures the property. Measurement errors might occur based on the differences from one appraiser's understanding of how to measure the property versus another's. Manual collection is not ideal for revaluation, as it makes it necessary for an appraiser to physically search for the property and property cards might not be organized according to the physical location of the properties. Paper maps could, therefore, be needed in order to locate the properties.

Figure 3.7: Field Mobile Data Collection Application. *Source:* Nate Maher, Appraiser Supervisor- Desktop Review, Maricopa County Assessor's Office

Another issue that could impact the quality of data collected includes the subjectivity of the assessor and other staff with respect to data coding. Coding of property data in a consistent manner is absolutely essential to the future development of accurate models and valuations. Care should be taken to train staff on how to identify proper attributes during collection. Data collection manuals are often created by jurisdictions which provide examples of attributes and their respective codes. When followed strictly, these manuals can be extremely helpful to assessors. For example, quality of construction is one of the most important variables that account for value, and assessors need to be trained so that they are able to identify the various conditions, and apply them with consistent accuracy.

3.2 | Electronic Field Review

Electronic field review involves using a mobile device in combination with GIS for locating properties while out in the field. Electronic devices can enable efficiency through primary digital data collection and the ability to reference existing data that might not be available during a typical field review. This is often performed on a mobile device or handheld that utilizes a field-friendly application for data collection.

The devices often provide a version of the CAMA system that has ready access to maps, documents, and sketching (*Figure 3.7*). This allows the assessor to conduct a full review of the property without having to write anything down on paper, and only changes the data as necessary. Sketching is done in the field, where a laser measuring device might be used to obtain measurements of the interior and exterior of the property.

3.3 | Drive Review

Drive reviews are often used for quality assurance purposes, to ensure the uniformity of new valuations. A drive review is just that—an assessor drives through a neighborhood and compares the assessed value, or other property data provided by either paper maps or field devices, of the properties. Parcel-to-parcel comparisons are generally made though subjective observation and knowledge of the market in the area. Drive review should supplement, but not replace, the other methodologies discussed. This methodology should be employed before and/or after new valuations are calculated in order to obtain the general environment and economic status of the neighborhood under review.

3.4 | Desktop "Office" Review

Desktop or "office" review combines several of the elements described in Section 2 for each of the data types that are maintained by assessors. This type of review requires access to the most up-to-date property data, imagery, and sketching for the property under review. This information is usually synced together by means of an integrated application so that assessors can review all of the information quickly and without having to search through multiple applications.

An advantage of desktop review is that it saves staff the time and energy they would have to expend if they were to go out into the field.

An advantage of desktop review is that it saves staff the time and energy they would have to expend if they were to go out into the field. It is simply a much faster and more efficient method of review. Desktop review procedures vary by jurisdiction, but when a review is well-planned it can provide similar appraisal performance as a field review, and use far fewer resources to complete.

Many offices are moving toward desktop reviews in order to accommodate growing tax bases and stagnant staffing. These types of reviews are most productive when data collection practices incorporate the guidelines set by the IAAO *Standard on Mass Appraisal of Real Property*, Section III. Section 4 of this chapter describes some best practices for conducting desktop reviews and expands on how one can incorporate them into property review workflows.

4 | Desktop "Office" Review Methodologies Integrating GIS

Revaluation reviews are arguably the most vital part of an assessor's function as they provide the data necessary for calculating property valuations. Errors in data collection often result in incorrect valuations. Additionally, if properties are not regularly reviewed, data become outdated, which also results in incorrect valuations. Since many jurisdictions have growing tax bases (as a result of development and stagnant office staffing due to budgetary constraints), desktop "office" reviews are an attractive alternative to field reviews.

However, it should again be noted that a jurisdiction may have statutory limitations on the use of office review as a valid alternative for collecting data that requires a physical inspection. Thus, the use of office review must be properly planned and approved by the appropriate oversight agency. A successful desktop "office" review program maintains several important components, including effective planning, a good workflow process, tools with which data can be appropriately captured, and a quality assurance program.

Many jurisdictions are seeing massive return on investment with respect to the use of

Figure 3.8: Example of Desktop Review Screens

desktop review for the revaluation of properties. Jurisdictions with a very large parcel base and/or a large land area have benefited tremendously from the use of this methodology, which allows them to allocate resources to where they are more needed.

4.1 | Planning for the Review

Planning for desktop review involves several steps as part of a full revaluation program. Most office reviews are planned during revaluation, when the areas that will undergo review are chosen. As with any program, a sales ratio study is conducted to determine which areas or neighborhoods would benefit from a potential review. The following things should be also known about a neighborhood in order to make this determination, among others: (1) whether the neighborhood contains sales; (2) how many basements are fully finished; and (3) whether there is any other evidence, such as might be found through building permits, to suggest that properties in that neighborhood have been updated.

Other things to take into account when deciding whether a specific property should be included in desktop review include the year that property was built, the date of the last physical inspection, and so on. *Figure 3.9* displays potential logic that can be used to determine whether a property is eligible for desktop review. Local knowledge of the market area should be utilized to make determinations about what should be included in a desktop review as opposed to a field review.

An active parcel with a dwelling record is eligible for desktop review if:

Single family residential use

and

Building count=1

and

No percentage complete on dwellings and additions

Table depreciation override on dwelling is blank

and

Legacy value method is not FIXED

and

Dwelling does not have economic or added physical depreciation.

and

If the Year Built is less/older than **1988** and no ownership changes have occurred since 01/01/2002 – skip the basement tests in italics below – parcel is eligible for deskreview.

If the Year Built is less/older than **1988** and there has been an ownership change since 01/01/2002, then perform the basements tests in italics below.

If Year Built is greater/newer than **1988**, then perform basement tests in italics below.

If Year Built is 2008 or greater and no inspection on record, skip the basement tests in italics below – parcel is not eligible for deskreview.

If Year Built is greater than 1988 and

If Dwelling has (Misc Codes=09, 10, or 11) and Basement Finish ratio > =30%

OR if dwelling does not have (Misc Codes <> 09, 10, or 11) and Basement Finish bsmtfin ratio > =50%

(If no basement subareas exist, skip basement test)

*****FINAL DETERMINATION OF ELIGIBILITY*****

Figure 3.9: Example of Desktop Review Logic. *Source:* Esri and its licensors; City of Fargo, ND , Esri, HERE, Garmin, Intermap, increment P Corp, GEBCO, USGS, FAO, NPS, NCRAN, GeoBase, IGN, Kadaster NL, Ordnance Survey, Esri Japan, METI, Esri China (Hong Kong), swisstopo, OpenStreetMap contributors, and the GIS User Community.

▶ 4.2 | Creation of a Workflow Process

Workflow processes are an essential element in any office review. Using a well-established CAMA system to manage the workflow of an office review allows for greater administrative control over the program, which helps to ensure that properties are being reviewed efficiently. Many workflow programs allow for the division of office reviews by staff, and guide them through the process of reviewing data for consistency. A good workflow program also gives staff the option of transitioning a review from desktop to field, if during the desktop review the assessor finds evidence that a field review is needed.

There may be other considerations built into a workflow that could be applied based on changes in property data. For example, a workflow might specify that a review should be sent to a supervisor for additional review if the change to the property's value is greater than a certain percentage.

▶ 4.3 | Tools for Data Collection

Desktop review integrates various data elements to create a "one-stop shop" for valuation review. A desktop review must be as efficient and accurate as possible, and there are several tools that integrate with CAMA to make this possible.

- **Change detection** is a process or algorithm that is conducted to determine whether property information in a particular area has changed over a specified period of time. This process manipulates aerial photography or satellite imagery in order to identify changes in development with a percentage of confidence. In desktop review, change detection allows an assessor to flag particular parcels that need to be reviewed more closely.

  For more information on change detection, see Case Study 3.1 ("Simple Change Detection Analysis") at the end of this chapter.

- **Imagery and remotely sensed data** provide the basis for desktop review, as imagery is what allows assessors to review the property remotely. This can include the use of: (1) aerial imagery taken over the course of multiple years; (2) oblique imagery, used to determine construction qualities or conditions, along with other types of wear; and (3) high-resolution, street-level images to confirm that what an assessor sees in the other types of imagery is correct. It is essential that whatever types of imagery are used the assessor is able to perform measurements and view the property from various angles.

- **Three-dimensional data** use LiDAR point clouds to create 3-D renderings of a property. Oblique imagery or street-level imagery can overlay these renderings to create a meaningful display. 3-D representations can also be used for line-of-sight analysis or for mapping condo characteristics and views. Integrating a 3-D property or 3-D cadastre with desktop review enhances the assessor's ability to analyze high-rise condominium views, and their ability to ascertain the amount of area and common space attributed to the value of the condo.

- **Sketching tools** are often used to create a scale drawing of the property in order to identify the quality of finish with respect to a particular area, such as a basement. Sketches are often produced using a sketching application that is integrated or built within the CAMA system, where vectors and angles can be specified based on primary measurement (i.e., measurement taken in the field) or imagery. Property sketches are important for identifying the typicality of a property's layout.

- **Multiple listing sites**, or other websites with property data, often provide additional useful information about properties under review. This is especially helpful in the case of desktop reviews, as it could help to identify whether there have been any updates to the property that might warrant a field review and that might not have been captured through building permits. These sites also provide useful sales information with respect to terms and conditions of the transaction that might need to be noted for additional follow-up. However, caution must be taken when using secondary or third-party sources, as the information they contain might not be accurate. Take care to cross-reference and follow up on questionable data to ensure that the information entered into CAMA and used to make decisions on valuations is adequate.

- **CAMA data** are the most important elements that must be integrated in a desktop review process as they are the primary means for storing data on the property. Data forms should be set up so that they are easy to edit and provide an orderly flow for the review process. The forms can also give the ability to add and edit land and property characteristics, information on building permits and sales, and the date and type of the most recent inspection.

As described in Chapter 2, the best way to use imagery within GIS is to integrate it with CAMA. In order for desktop review to work as it should, it is essential that CAMA is well-integrated with a GIS application so that it produces the correct imagery and data layers. In addition, well-integrated sketching programs will be necessary to allow the user to move from parcel to parcel without needing to have multiple applications open at the same time.

In order for desktop review to work as it should, **it is essential that CAMA is well-integrated with a GIS application** so that it produces the correct imagery and data layers.

See Case Study 3.2 ("The Desktop Revolution: Changing the Fabric of Canvassing in the Assessment Community") at the end of this chapter for a real world example of how one county made exceptional use of desktop reviews.

4.4 | Quality Control

When the tools and data elements are used together, they form the basis upon which decisions about properties can be made in a matter of minutes. However, if the data are not high-quality or a mistake is made somewhere in the process, the resulting valuation might not be accurate. The following measures should be taken to ensure the accuracy and consistency of desktop review methods:

- Information should be reconciled through imagery-based review.
- Control measures should be integrated throughout the workflow to guarantee that if a major change is made to the data—such as a difference in quality grade or excess square footage—a second set of eyes (preferably a supervisor) is called in to review and approve the changes.
- The property data should be mapped in order to analyze the spatial consistency of the data.
- Office reviews should be supplemented with a drive review of the neighborhood after the valuations are calculated to ensure that the site is consistent with the findings of the desktop review. If, however, a jurisdiction does not have the resources to devote to conducting an onsite review of any kind (which might often be the case), it is even more essential that they use the most up-to-date imagery to their desktop reviews.

Other data quality considerations that are typically found in CAMA systems can also be applied to desktop review, including range edits and the identification of missing data and invalid coding. The next section will discuss the preservation of data quality in much greater detail.

5 | Data Quality and Verification

Providing a visual depiction of the spatial distribution of property data has been one of the most common uses of GIS in the field of property valuation. Many jurisdictions are able to create a data extract containing CAMA and sales data and attach it to the parcel geometry. Symbols and/or colors are then added to this geometry to represent different characteristics. This gives assessors the ability to look for consistency within data at the local level, instead of in aggregate, which provides additional context to the conformity of the neighborhood under study.

5.1 | Mapping Continuous Property Data

Data are continuous if they can be measured on a numerical scale, and can take on any value (including decimals). Assessors often map continuous data—which are displayed in various colors, based on a defined classification—in order to discover anomalies within a specific region. A **classification** is the proces of storing or arranging entities into groups, bins, or categories. When mapped, each of these groups is assigned a different color so it can be easily distinguished. A map legend should include a key that describes which classification each color represents. *Figure 3.10* shows an example of continuous data shown on a map.

There are many different methods of data classification that can be used to break the data into different bins or groupings depending on the distribution that will best represent the data. Some of the most common data classification methods are described in *Table 3.6*.

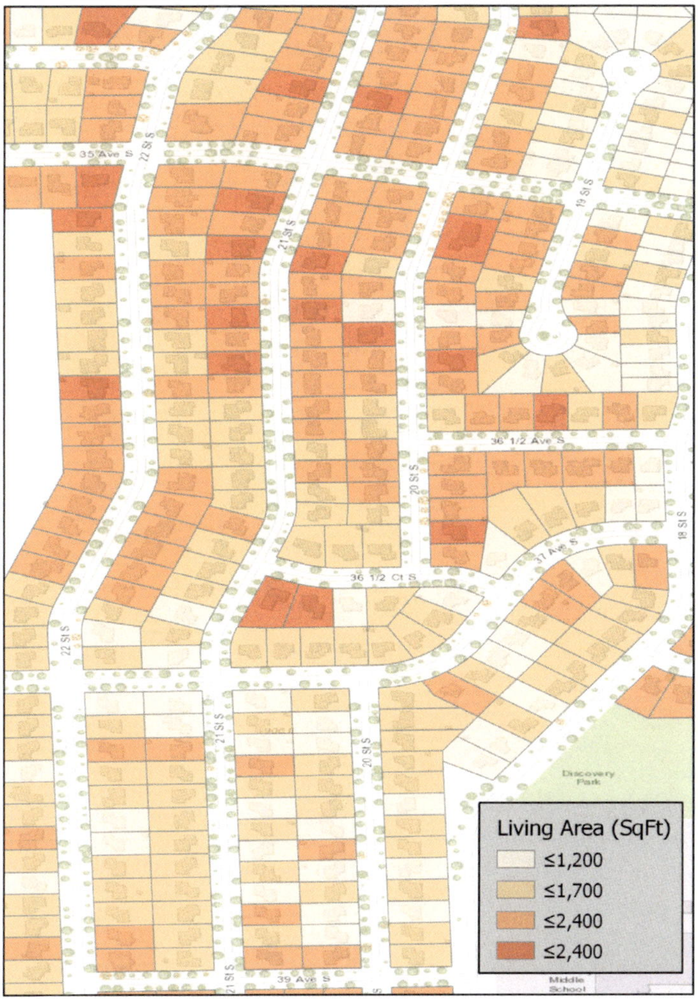

Figure 3.10: Continuous Data Example Using Living Area on Parcel Polygons. *Source:* Esri and its licensors; City of Fargo, ND , Esri, HERE, Garmin, Intermap, increment P Corp, GEBCO, USGS, FAO, NPS, NCRAN, GeoBase, IGN, Kadaster NL, Ordnance Survey, Esri Japan, METI, Esri China (Hong Kong), swisstopo, OpenStreetMap contributors, and the GIS User Community.

Table 3.6: Common Data Classification Methods (ESRI, 2017)

CLASSIFICATION TYPE	DESCRIPTION
Manual Interval	Defines your own classification scheme
Defined Interval	Allows specification of interval size based on number of classes within same value range
Equal Interval	Classes are divided out equally based on defined number of classes
Quantile	Each class contains an equal number of data values
Natural Breaks (Jenks)	Based on natural groupings within the data (maximizes the differences between classes)
Geometrical Interval	Based on an algorithm that minimizes the sum of squares for the elements in each class
Standard Deviation	Breaks based on standard deviations from the mean

It should be noted, however, that classifications also introduce several biases that may mislead interpreters of the map. Mark Monmonier describes many of these biases and misconceptions in his book *How to Lie with Maps* (1991). The generalization of data often presents problems with respect to using standardized data as opposed to raw data (a phenomenon referred to by Monmonier as "lying with standardization") or with which classification is used based on the distribution of the data (or, "lying with classification"). These "lies" can often distort how colors or symbols represent the data on the map and covey a meaningless or completely false message to the interpreter. These are issues that many assessment professionals may not necessarily consider when creating maps. There are several excellent books on mapping that address these issues, including two by Cynthia Brewer. Information about them can be found in the appendix under "Additional GIS Resources."

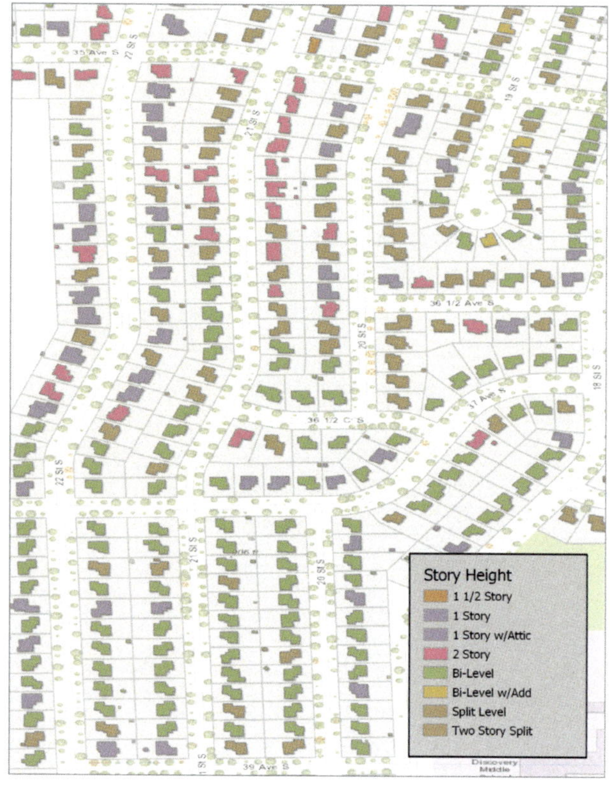

Figure 3.11: Discrete Data Example Using Story Height on Building Footprints. *Source:* Esri and its licensors; City of Fargo, ND , Esri, HERE, Garmin, Intermap, increment P Corp, GEBCO, USGS, FAO, NPS, NCRAN, GeoBase, IGN, Kadaster NL, Ordnance Survey, Esri Japan, METI, Esri China (Hong Kong), swisstopo, OpenStreetMap contributors, and the GIS User Community.

5.2 | Mapping Discrete Property Data

Discrete data also referred to as **categorical data**, are often associated with a particular object that has defined boundaries. These types of data are very common in the assessment profession as they describe quality of construction, property type, story height, type of siding, and so on, which are in turn used to describe a property on a map. Discrete property data are often visualized with a single color or symbol associated with a specific feature type. This could also be denoted through the symbol's size and/or type. *Figure 3.11* shows an example of how categorical data might appear on a map.

5.3 | Discovering Missing Data, Incomplete Data, and Outliers

One of the benefits of using GIS for data quality assurance is that because the property data are represented visually, it is easy to spot outliers. However, decisions made using data visualization are only as accurate as the data collected through primary or secondary means. Therefore, data that are collected through secondary sources or from a third party should be verified and cross-referenced against other sources.

Missing data are often easily discoverable through visualization on the map since they do not have any associated coding scheme with other displayed data. Outliers are found by analyzing coded data that do not conform in a particular area, or by discovering a particular classification of data whose range also does not conform to the surrounding area. GIS enhances visualization for assessors by providing this information for an entire jurisdiction, which can be updated as data are updated within the CAMA system. This makes it ideal for office data review programs.

5.4 | Visualizing Patterns and Aggregations

Visualizing property data in aggregations or groupings based on meaningful boundaries—typically based

The modifiable areal unit problem (MAUP) is a common geographical issue with respect to continuous data and the generalization of spatial analysis. MAUP is a "problem arising for the imposition of artificial units of spatial reporting on continuous geographical phenomena, resulting in the generation of artificial spatial patterns" (Heywood, 1988). There are two distinct types of MAUP:

Scale MAUP. The spatial scale at which information is analyzed often determines the generalization of phenomena. An example might be vacancy rates of apartments at the neighborhood level denoted by an assessor as opposed to at the jurisdiction level. The scale at which a phenomenon is analyzed often affects how data are interpreted, thus impacting the decisions that arise from those interpretations. Therefore, it is always recommended by geographers that data aggregations have meaning within the results of the analysis.

Zone MAUP. The grouping of phenomena (density) are never dispersed equally—thus, potential zones or shapes created due to clustering of phenomena are never in equal zones. In other words, there are major differences in the aggregation of property values depending on how neighborhoods are divided and separated, no matter the scale.

on hierarchical units at the neighborhood, group, or jurisdiction level—is not only quite aesthetically appealing, but also extremely helpful to assessors. Aggregating data provide summary information for the unit under study. Ratio studies are often calculated and visualized at the neighborhood level (or some larger, aggregate level) from a geographical perspective, which then allow assessors to make decisions, based on appraisal performance and equity measures.

Being able to analyze the spatial distribution of sales is also important to determining market activity within a neighborhood. There are many spatial statistics that can be used to quantify spatial patterns for clustering, calculating spatial regression models, and looking at density. However, care must be taken when analyzing aggregated data. If an assessor is not careful, they can fall prey to the modifiable areal unit problem and mislead those who later try to interpret the analysis. This problem is defined in the following box, along with two subtypes of this issue.

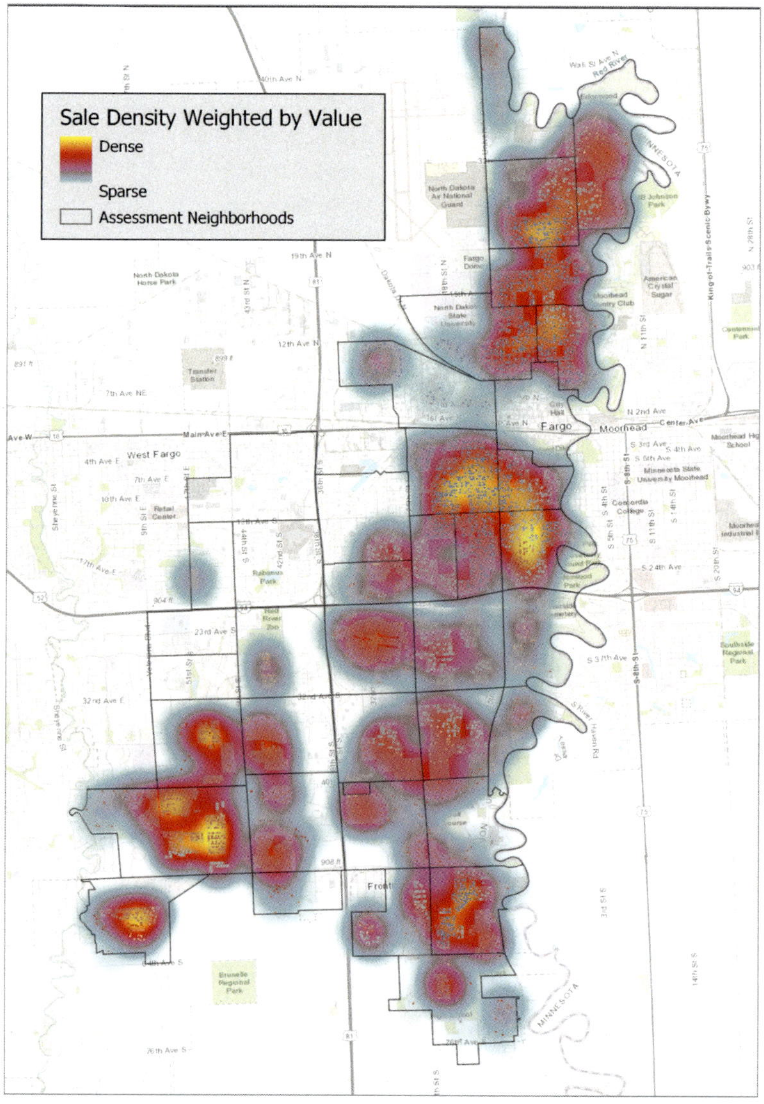

Figure 3.12: Map Showing a Surface of Sales Density Weighted by Modeled Values

6 | Conclusion

Throughout this chapter, we sought to provide context for the efficient collection of accurate property data. Assessors collect data from a wide array of general, specific, and comparative data sources in order to enhance the calculation of fair and equitable property valuations. Data are often collected on physical, economic, governmental, and social factors. Physical factors—such as the location of properties relative to roads, rivers, lakes, commercial property, and other features—help assessors understand the relative location of a property. Economic data, such as rates, rents, and economic measures, are collected in order to understand the quality of neighborhood a property resides in. Governmental data with respect to regulations often influence where and how properties are built. Social factors, such as population and crime rates, often determine the demographic makeup of neighborhoods.

Data collection has evolved through a wide variety of methodologies, all of which are still used in many jurisdictions around the world. To review, these methods are:

- **Manual collection,** where data are collected in the field
- **Electronic collection,** which involves collection via field mobile devices
- **Drive review,** which is a quality assurance methodology
- **Desktop or "office" review,** where imagery and other secondary or third-party sources are used to collect information on a property

Desktop review itself can be broken down into various methodologies, although all involve conducting property reviews in-house rather than going out into the field. This requires the integration of several programs, many of which involve the use of GIS and other spatial data as a medium for visualizing location.

Using GIS as the primary integration piece has many advantages. First, it provides a visual display of the work that has been completed and the work that still remains to be done in a review. Then, it allows for the visualization of property data to ensure that data are being coded consistently. Finally, GIS allows for the use of imagery, sketching, and other tools to be displayed and overlaid at the location of properties, which gives assessors the ability to analyze other neighborhood characteristics. A desk review program will follow several important processes, including a planning stage, workflow creation, the integration of essential data collection tools, and many levels of quality assurance.

Data quality and the verification of data are essential to the consistency and accuracy of data collection. The ability to view and analyze these data spatially is paramount to assessors who seek to identify conformity issues or anomalies within specific neighborhoods. The use of more advanced spatial statistics, such as clustering, help quantify the degree of spatial patterns and can often indicate the reason why a particular phenomenon exists within the data (e.g., what is influencing sale values within a particular neighborhood). Overlay and more advanced statistical functions are the focus of Chapters 4 and 5.

Case Study 3.1

Simple Change Detection Analysis

Author: John Watterson, GISP • Martin County Property Appraiser's Office, Florida

Change detection may be considered an application of a geographic field parallel to GIS. Remote sensing. remote sensing is defined as the collection and interpretation of information about the surface of the earth from a distance from the earth (1). Two of the most widely used remoting sensing products available are aerial photography and satellite imagery. Aerial photography is captured by a type of aircraft, such as a plane or a drone. Change detection often involves the comparison of two aerial photographs or images of the same area at different times (2). Software currently exists that can perform an analysis on two sets of photography or imagery, and produce a third image which can be manually adjusted to highlight areas of possible change. In the property assessment industry, change detection is mostly used to discover some type of difference in the building footprint of a parcel, such as an addition. Other types of improvements that may be identified are pools, sheds and separate structures. Other, less common, uses of change detection for property assessors include crop monitoring and vacant land analysis.

For the purpose of this case study, we will focus on the use of a simple, readily available GIS tool to perform change detection using aerial photography. The tool can be used to find significant change, which can lead to the recovery of missing revenue.

The name of this tool is Image Analysis and it is available as a button on the default "Tools" toolbar in ArcGIS Desktop version 10.x. The tool has various windows to adjust settings to perform the actual change detection. The top window allows for the loading of different time periods of imagery, and is also where the final change detection image is created and added. There are also "Display", "Processing" and "Mensuration" windows within the image analysis tool that allow for the setting of various parameters, and the image processing itself.

The property assessor's office in this case study has had considerable success using the Image Analysis tool. Over several years, the assessor's office has repeatedly found missing improvements, which were then forwarded to residential and commercial valuation departments for verification. Many of the missing improvements never received an official permit for construction, and therefore represented missing value to be recouped by the tax assessment roll. These unpermitted improvements occur yearly, both in urban and rural areas, in the moderately populated county in which this case study is based. In the past, these improvements were almost impossible to detect and verify by the traditional field check, especially due to property access issues, mostly in rural areas. A significant benefit of this tool is that not only has it been used to find missing revenue, but the tool itself is of no additional cost; it comes pre-loaded with ArcGIS Desktop 10.x. At the assessor's office in which this case study is based, this process is accomplished annually, since new aerial photography is received each year. Utilizing this tool, the GIS professional essentially has a bird's eye view to see change.

The process of creating the necessary raster (pixel-based) data set to visually inspect for unpermitted improvements is straightforward for the GIS professional.

Before launching the Image Analysis tool, two different years of aerial photography are required to be loaded into the Table of Contents in ArcMap. The years do not necessarily need to be consecutive. When the Image Analysis tool is launched, all years of imagery in the Table of Contents will automatically be added to the top window of the Image Analysis tool. One should only work with two years of aerial photography when using the tool. The next step is to select/highlight the two aerial photography layers in the top Image Analysis window. The older aerial photography should appear above the most recent year. Next, the "Difference" button should be clicked in the Processing window of the Image Analysis tool to produce a new raster layer that can have its color adjusted to show darker areas, which indicate some type of change. Now, the new raster layer created (named by default: Diff_image), should be highlighted in the top window of the image analysis tool and have its Gramma value adjusted by using the Gamma slider, which is in the Display Window of the Image Analysis tool, to a considerable higher value than the default, preferably 6.03. The term Gamma refers to the degree of contrast between colors in a raster data set (3).

This setting will colorize the new raster layer so that significant change will appear darker, standing out from the rest of the image, which will be in light color. Sometimes these dark areas turn out to be false positives, such vegetation change, but often indicate some type of new construction. The next step is to create a shapefile or geodatabase feature class of parcels that have received an official permit for some type of construction. This can be done by joining a spreadsheet of permits to a parcel layer, and then selecting out parcels that joined to the permits spreadsheet. Then, the selected records in the parcel layer should be exported as separate permit layer. The permit layer should be loaded into ArcMap, along with the countywide parcel layer. The permit layer's symbology should be set to a solid color and the symbology of the county-wide parcels should be hollow.

The permit parcel layer will now cover the dark areas where authorized construction has occurred, while the hollow county-wide parcel layer will show dark, unpermitted change, if the area is not already light. When dark areas are discovered that seem to resemble construction, the GIS professional would then turn off the change detection raster layer, and then visually compare the two years of imagery by turning them on and off in the Table of Contents of ArcMap to see and judge the change. For parcels that do appear to have a questionable improvement, another separate parcel layer should be created by selecting and exporting those parcels so they can be tracked. Once the entire county has been scanned, the last step is to create a spreadsheet of Parcel Identification Numbers from the questionable parcels GIS layer and then provide it to residential and commercial appraisal departments for review using their appraiser-level GIS software. They will then determine if the change is worth researching.

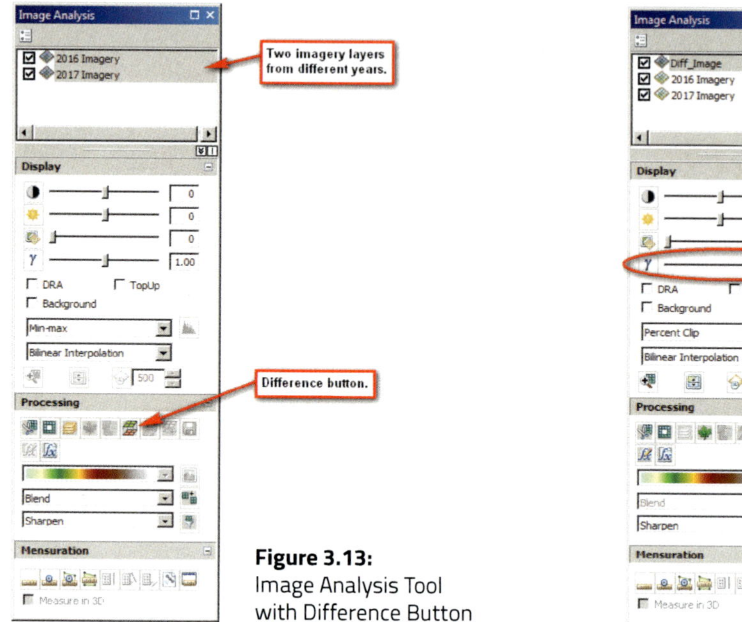

Figure 3.13:
Image Analysis Tool with Difference Button

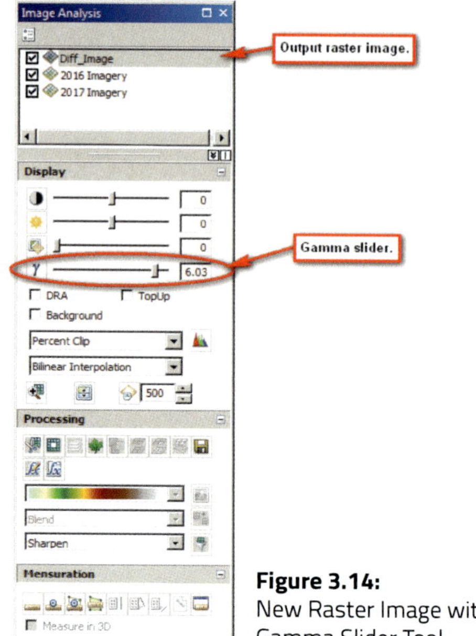

Figure 3.14:
New Raster Image with Gamma Slider Tool

Figure 3.13 shows the Image Analysis tool with imagery years and the Difference button, which are used in the first part of processing.

Figure 3.14 shows the new raster image that is created from the initial processing and the Gamma slider tool, which is used to adjust the color of the raster image for change detection purposes.

Figure 3.15 shows the image analysis output raster layer on the left image with a darker area on a highlighted parcel, indicating an area that has possibly changed. The image in the center shows an aerial photograph from the late year, with no improvement in the questionable area. The image to the right shows an aerial photograph from the recent year with a pool showing. The two years of aerial photography were used to create the raster image on the left. In this example, it is clear that change has occurred that warrants investigation.

Figure 3.16 is a portion of the final spreadsheet provided to appraisal departments containing parcel records with identifiers, as well as a comment column populated by the GIS professional. These parcels are reviewed for significant change.

CITATIONS

1. Wade, Tasha and Sommer, Shelly (2006) "A to Z GIS" (p. 179), Redlands, CA: ESRI Press.
2. Wade, Tasha and Sommer, Shelly (2006) "A to Z GIS" (p. 30), Redlands, CA: ESRI Press.
3. Esri (1995-2015) "ArcGIS 10.3.1 Help"(Stretch function)

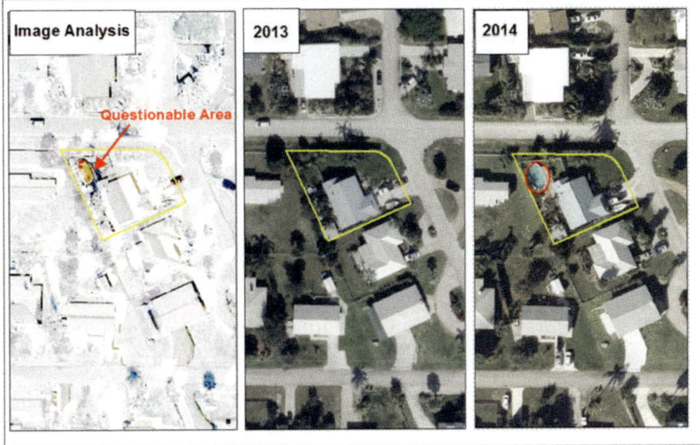

Figure 3.15:
Image Analysis Output Raster Layer (left); Aerial Photograph from Late Year (center); Aerial Photograph from Recent Year (right)

PCN	AIN	COMMENT
20374100700000002000000	196831	new building
15374100400040005010000	690	new building
28374101400000050000000	724921	new building
28374101400000054000000	390903	new building
28374101400000052000000	547322	new building
30374201500000014050000	10273	construction area
30374201500000015050000	10274	construction area
04384103500000001000000	1116208	new construction
04384103500000002000000	1116209	new construction
04384103500000003000000	1116210	new construction
04384103500000004000000	1116211	new construction
04384103500000005000000	1116212	new construction
04384100100500003080000	21252	demo
05384100501800004000000	1115775	new building
16374100300400001090000	1345	demo
26374100500000127000000	149976	new construction
26374100500000137000000	105956	new construction
38341027000000001000000	1115829	new construction
38341027000000002000000	1115830	new construction
49384100100100018190000	39353	new home
55384100006600008110000	44270	new building
55384100800000010090000	44475	new building
30384200404400017070000	47240	new structure

Figure 3.16: Final Spreadsheet (Portion)

Case Study 3.2

The Desktop Revolution: Changing the Fabric of Canvassing in the Assessment Community

Author: Nate Maher, Appraiser Supervisor • Desktop Review, Maricopa County Assessor's Office

In 2013, Maricopa County set an ambitious goal: to canvass 1.37 million residential parcels in ten years utilizing desktop review. Maricopa County is the fourth most populous county in the United States, valuing nearly 1.7 million real and personal property parcels/accounts with a full cash value of more than $508 billion in 2018. In a county of 9,224 square miles, daily high temperatures reach 100F (38C) or more for thirty percent of the year.

The most efficient approach was to conduct a desktop review of the county. This was done utilizing a GIS platform to store, analyze, and manipulate canvassing data. Maricopa County acquired Esri Canada's Assessment Analyst, based on the ArcGIS platform, to serve as the foundation for their review. EagleView delivered orthophotos and obliques, and Cyclomedia prepared the corresponding street front imagery. Strict adherence was given to the International Association of Assessing Officers' Technical Standard 6.4, "Alternative to Periodic On-Site Inspections."
The county is now halfway through its fifth phase of desktop review, with over 325,000 completed parcels updated in the CAMA system. They have generated over $1.6 billion in additional full cash value (market), and over $1.7 billion in limited property value (taxable).

OBJECTIVES

The primary objectives of desktop review are to correct incomplete data for single-family residences in the Valley, and to adhere to international standards on canvassing. At inception, this was the largest desktop review in the country (if not the world) and remains so today. The first two phases included approximately 50,000 parcels, with each subsequent phase at 100,000 parcels.

In order to be successful within international standards and state statutes, we needed a system that would combine rule sets with multiple vendors and software platforms. We required a way of analyzing, storing, and manipulating data that would produce the most accurate and important product our offices provide the public: fair and equitable valuation. The natural conclusion: GIS was the conduit of choice to maintain accuracy, consistency, retrieval and storage of updated data in the most beneficial way to our offices, and our constituents. Further, utilizing GIS gave Maricopa County the most feasible platform to create, maintain, study, and render opinion of data. Desktop review was the catalyst for this massive update of data, and GIS was the county's most reliable tool of choice.

Adhering to IAAO standards was successful due in part to the relationship with vendors Esri Canada, EagleView, and Cyclomedia. All oblique and street front photography was captured within two years of the canvass, the technical standard for a rapid-growth jurisdiction. If foliage, poor photography, personal property, or other opinions led the appraisers to incomplete findings, they escalated parcels to the field for a traditional review. Fieldwork amounts to approximately 1% of the parcels within every desktop review phase.

Any property owner who received a substantial change in limited property (taxable) due to a "B rule" (which refers to the fact that in the state of Arizona, an addition to a parcel's improvements can trigger a recalculation of the base limited property value if the size or value of the addition exceeds certain thresholds), was sent a letter informing them of the specific change, and their appeal rights. The letter was sent in advance of the close of the valuation roll, allowing property owners to call in, request a recheck of their parcel, and for appraisers to make any changes due to an error by desktop review, prior to the next formal notice of value.

This letter gave desktop review appraisers an opportunity to educate the public on what changes were detected, how long they had gone without being on the roll, and for the public to correct us on a misunderstanding of the component's use (i.e. storage area attached to home, not a livable addition). Some of the escaped components were the doing of previous owners many times over, and it was important to get the current owners educated to the change in their valuation due to structures they themselves may not have added. This letter started the conversation with the public, welcoming their concerns or misgivings, and the office used it as a way to discuss new technology in assessment in an open and transparent way.

ANTIQUATED DATA

The desktop review process initially worked some of the oldest residential areas of our county. Many permitted and non-permitted improvements were discovered. Whether the county, municipality, or owners were to blame for their escape, these changes were picked up going forward for the valuation roll.

The prior state of initial data meant there were vast changes on most parcels. This required re-drawing sketches, researching through historical aerials the construction years of the improvements, and trying to determine quality of the escaped components many years later. This slowed down the typical timeframe for appraisers to complete a desktop review of a parcel. The average time to complete a parcel is about 15 minutes, which is still significantly faster than a field visit, including travel time. As this process enters newer areas of the county, or re-canvasses prior desktop review phases, we predict an uptick in completion time.

CONSISTENT CANVASSING METHODOLOGY AND BUY-IN

In Phase I of DTR, the county had a dedicated canvass team, created with new hires, and appraisers transferred from other crews. The initial concept of desktop review was met with resistance. Appraisers had long relied on field visits to ascertain changes in residential properties. They doubted the ability to accurately identify, measure, and update property records through aerials and street front imagery alone.

New practices had to mesh with current policies and procedures, so that the desktop review would insert accurate data into the property record. This included creating an initial desktop review manual by founding canvass supervisor, Tironda Dixon. It was modified by Nate Maher in 2015, to come into line with the residential manual and practices. As a result, the residential field canvassing that occurs in permit work now follows the desktop review practices. Both desktop and field appraisers are on the same page for updating residential parcels. This strengthens our data, and keeps our practices fair and equitable to property owners.

Buy-in to any major change takes time. It involves conversations with affected parties, as well as weighing what changes are superior, and which older practices require preservation. Over the course of several phases, Maricopa County continuously improves upon their process by de-compressing with all affected partners (appraisal division, administration, IT, and CAMA market modelers), adding what does work to the recipe, and removing the sour notes. As we move for-ward with a new CAMA system, our desktop review process will continue to grow in accuracy, and eliminate some latency in our transfer of data into our production database.

METRICS

Desktop review is unique in our office for its ability to track the time necessary to complete its core function: canvassing. Through the ArcGIS platform, Assessment Analyst, along with Micro-soft Access reports, our operational planning can be refined based on analysis of past phases. The software is able to follow the start and stop times of desktop review appraisers as they work their way through a parcel's completion. Over time, this performance data allows for a larger analysis of what the appraisal division will need in labor, time, and training to finish the phases on time. It also allows the desktop review supervisor to assist the administration in proposing future phases that can be certain of timely completion.

Over four phases, supervisors and management adopted increasingly challenging quotas. The current requirement for each desktop review appraiser is 625 parcels per month, or ~32 parcels a day, ~4 per hour. As appraisers become fluent in the practice of desktop review, we see some su-per users emerge (1,000+ parcels a month with an error rate of 5% or less). Currently, 15% of the staff are super users. Strong appraisers on staff lessen the urgency to make up for time off from appraisers, turnover, and technical difficulties. Their extraordinary work is noted in their perfor-mance reviews, and they are given additional projects and opportunities to train to develop their future leadership skills. On each iteration of desktop review over four phases, there have been at least two super users on every team.

A concern of appraisers and supervisors in the past is the monotony of the work. Desktop review is a predictable workflow with few seasonal changes (letters to public resulting in rechecks, end of year error reports, etc.). This allows for a strong, consistent work product (the past two phases of desktop review have finished two weeks ahead of schedule). However, it is up to the supervisor and his leadership to keep appraisers stimulated, learning, and in a position to continue to grow. One way Maricopa County has motivated their appraisers is through the international designations program with IAAO. The current desktop review team has two appraisers starting their AAS designations, and several others interested in the process. Another way the supervisor motivates staff is through team meetings, annual luncheons, access to conferences and non-IAAO learning (Appraisal Institute, Urban Land Institute), and submersion into other divisions in the assessor's office.

PUBLIC RELATIONS

When a new technology inserts itself into a current governmental process, oftentimes it is met with skepticism by its most affected parties: the public. A keystone of our desktop review at Maricopa County was transparency with the public. The more notice we could give property owners, and educational opportunities we could provide, the more likely desktop review could be seen as a positive data cleaner, instead of an inaccurate technology, erroneously adding value to parcels without cause, or even worse, with an arbitrary intent of increasing taxes.

This was not an easy conversion of public opinion. It involved skepticism as doubt was seeded in our ability to measure through "satellite imagery" (our aerials are from aircraft flying at 4,000 feet). It added appeals in a traditionally non-litigious valuation cycle, and increased other administrative valuation/component relief avenues (notices of claim, rechecks). The Arizona State Board of Equalization had members questioning the use of aerials to adequately value property.
In order to work through the misconceptions by the public we served, and the oversight organizations we answered to, the Maricopa County assessor's office became as transparent as possible. We enhanced the frequently asked questions on our website to describe the desktop review process, called all inquiring property owners on desktop review back within two business days, and always offered to have an appraiser physically come out to the home.

With our oversight agencies, the State Board of Equalization (SBOE) and the Arizona Department of Revenue (AZDOR), education and explanation was crucial to our program's success. These involved sessions with the chief deputy assessor, chief appraiser, project manager, and canvass supervisor educating the agency's representatives to the practices and procedures of desktop review. Our alignment with IAAO standards, along with Arizona statutes, made the explanation of our work flow better understood by all oversight parties.

The agencies learned our desktop review teams used measurable aerials, updated street photography, cross referenced active building permits, parcel histories, and our well preserved subdivision files. They discovered our quality control process was an in-depth, manual process of the supervisor reviewing their appraiser's work, building up Access reports to predict errors, and that the process still required field work when data was inconclusive on the desktop. Agencies heard of our letters to any owner with a significant change. They were walked through our desktop review recheck process. The conclusion of our discussions with these agencies showed that Maricopa County's desktop review team was operating in a deliberate, fair, and equitable manner to every parcel we were tasked with canvassing. That consistent treatment gave strength to the program despite misconceptions of technology's role.

SCALING DTR TO ANY JURISDICTION

So how might others scale desktop review to fit in a smaller jurisdiction? We suggest starting with IAAO technical standards on general canvassing and desktop review. Dependent on your jurisdictions statutory requirements for re-canvassing, one should build a review process that centers around a 4-6 year re-canvass of every parcel. Utilization of a vendor or inter-government collaboration to obtain aerial and street front photography will be most helpful in acquiring accurate desktop data. If you can reliably measure from aerials, you give your jurisdiction the best chance at success in desktop review. Without a measuring tool, you risk subjective estimations instead of a reliable replacement to a field visit.

Many jurisdictions employ a desktop review, some using an ArcGIS platform like Maricopa County, some developing their own wrapper to host multiple vendors' data, and some just pull from many areas (CAMA system, aerials, GIS maps, MLS listings) and make their updates manually. Each is a satisfactory way to update records with desktop review.

CONCLUSION

Ultimately, desktop review is a computer in an office instead of a clipboard in the field. It is a measuring tool on an aerial, not a measuring tape on a stucco wall. It is working finger strokes away from MLS listings, permit databases, parcel histories, and one staircase from some of the most well-preserved residential subdivision files in the country dating back to the 1970s (files with original blueprints, sales brochures, and extensive notations from appraisers anywhere from 1 to 40 years ago). All those pivotal data sources are currently unavailable in the field. Instead of the immediate research desktop review promotes, field appraisers scribble notes on a field card, hopefully to be investigated once they returned to the office. Desktop review eliminated not only travel time to parcels, but gave immediate access to these databases to conduct the investigatory work required on appraisers in making an opinion.

Desktop review adds a collaborative culture missing in the independent field appraiser's role. If a desktop review appraiser has a question on a component's use, she has 13 other appraisers (and a supervisor) within feet of her work space. There is an active learning environment between new hires and super users, and a leveraging of each appraiser's strengths as we work through residential properties ranging in size of 500 to over 15,000 square feet. Team meetings give an arena for solving new found errors, and sharing short cuts and precision moves created by other appraisers. The desktop review supervisor works laterally with the CAMA and information technology divisions, chief appraiser, and administration to build a better, more accurate product for our citizens going forward. Desktop review is one way an assessor's office can get out in front of change, for the betterment of their data, and with cost savings to their constituency in mind.
GIS drives the consistency and successful communication of multiple resources (ArcGIS, Assessment Analyst, photography, measuring tools, and appraisal staff) to make desktop review a viable tool for canvassing residential properties. We live in a world demanding dense analysis of data from myriads of areas. Assessment is no different. The greater fluency our data analysis has, the better chance assessors have to produce a fair and equitable value for our constituents.

Case Study 3.3

Street-Level Imagery Case Studies in Communities Small, Medium, and Large

Author: Jennifer Kuntz • CycloMedia Technology, Inc.

Assessors in counties and cities of all sizes benefit from technology to make efficient use of staff and funding using the alternative to periodic on-site inspection from the IAAO *Standard on Mass Appraisal of Real Property*. Section 3.3.5 of the standard states that "Current high-resolution street-view images (at a sub-inch pixel resolution that enables quality grade and physical condition to be verified)" should be among the datasets included for desktop review.
Assessors have been using street-level imagery for decades, as illustrated with this vintage photo including the, at the time, state-of-the-art method for adding a parcel identification number.

The evolution of cameras and digital processing has taken us from the creative parcel identifier labeling in our vintage photo to digital images with sub-inch pixel resolutions, optionally tagged digitally with a parcel identification number and date. Newer images are delivered in a digital format, such as JPEG, and were, until recently, typically a rectangular image that captured the property in question with sufficient detail to determine grade and condition.

Figure 3.17. Photo Courtesy of Marlene Jeffers, CourthouseUSA, formerly Mobile Video Services.

The next generation of street-level imagery and associated software tools can provide more than a rectangular high-resolution picture of condition and grade. Professional-grade street-level imagery, which can be captured by a vehicle driving at normal traffic speeds along roadways and alleys, can capture a series of panoramic images that allow the viewer to see condition and grade as well as take measurements of structures. Because the images are taken in series, typically several meters apart, the viewer has the option to see the property from multiple angles and to view the surrounding area.

The following characteristics are typical of the next generation of professional-grade street-level imagery:

- 360-degree panoramic images in a regularly-spaced series along streets and alleys to allow views of properties from multiple angles

- Pixel sizes of a fraction of an inch to provide detailed views of building materials and property condition

Figure 3.18: Confirming width of building shown on a building sketch using street-level imagery to measure under the eaves. Screenshots Courtesy of Clarence E. Mingo, II, Franklin County Auditor and CycloMedia.

- Tools to capture rectangular photographic cut-out images of areas of interest for use on property cards or other reports
- High positional accuracy that will allow for GIS feature data collection directly on the imagery
- Ability to measure accurately from the imagery
- An option to host on the cloud for convenient access

End users should also look for their imagery vendor to provide an application programming interface (API) for integration into CAMA, desktop review, or other systems. Availability of an API will allow developers to more easily integrate the imagery and associated tools into existing systems or to create a customized user interface to meet specific workflow needs. Another question to ask is whether the vendor already has integrations into systems currently in use at the user site.

The positional accuracy, the ability to measure on the imagery, and the view of surrounding conditions offer some valuable enhancements over a rectangular snapshot. The panoramic images can be used to confirm building sketch measurements that are not visible from the air such as the width of a building under the eaves. Because the images show the surrounding properties and neighborhood factors, assessors can view nearby conditions that will influence value, such as external obsolescence factors like a highway overpass or commercial development or positive influences such as a park or a lake view. Either the panoramic images or photographic cut-outs made from them can be used to assist an appeals board in determining the legitimacy of a property-owner's appeal.

The imagery can be a community-wide asset, allowing other local government departments to benefit from viewing conditions in the field, collecting data, and taking measurements. For example, a public works department can collect hydrant locations or improve the positional accuracy of a street light database. Next Generation 911 address locations can be captured at the front door of buildings. Elevations at grade and at the first floor of living space within a structure can be captured to understand the impacts of potential flooding.

The following case studies describe the use of street-level imagery in jurisdictions with a wide range of parcel counts and a variety of systems with which they have integrated the imagery

MULTIPLE USES AT MASON CITY

The assessor for the City of Mason City, Iowa, home of approximately 14,000 parcels, acquired street-level imagery in cooperation with the GIS department at the City. The GIS department was interested in the acquisition to collect locations of infrastructure assets such as signs and streetlights. Assessor Dana Shipley saw the benefit the imagery could bring to her department as well. The arrival of the imagery was timely for staff to update their database with photographic cut-out images for their commercial properties. Mason City contracted with the vendor that makes their property information website to add the 360-degree image views into the map viewing portion of their application. The integration of the imagery with the existing application makes it convenient for Mason City staff to verify and update the property information in their database. Users can click on a building or parcel on the map to view the imagery at that location. With the image open, the user can navigate to nearby images to view the property from a different angle, pan and zoom to see details, and download a photographic cut-out of the current view. City staff can also take measurements to confirm building dimensions.

CUT-OUTS AND CAMA AT DAKOTA COUNTY

Dakota County, MInnesota, with approximately 160,000 parcels, was interested in updating all records in their CAMA database with new imagery to be used for property cards and their website. They chose to have street-level imagery collected with photographic cut-outs created by the vendor and supplied to the County with tabular data that allowed them to insert the images into the appropriate records in their CAMA database. The CAMA vendor used by Dakota County integrated the panoramic images with the CAMA system, providing Dakota County's appraisers with a window that lets them view property details and measure building dimensions at the street level inside their CAMA interface.

MARICOPA DESKTOP REVIEW

Maricopa County, Arizona, with approximately 1.6 million parcels, uses a desktop review system that integrates the components listed in section 3.3.5 of the IAAO standard tightly with GIS. A unified interface lets appraisers view GIS data, obliques, building sketches, status of appraisal projects as well as street-level imagery. As in the other communities, the appraisers can use the street-level imagery to view grade and condition of structures and take measurements that might not be visible from other views such as dimensions under eaves or overhangs. "We have used street-level imagery provided by CycloMedia with our Assessment Analyst© software tool to systematically update property data and capture information that was missing or inaccurate in the property tax roll." Said Tim Boncoskey, Chief Deputy Assessor for Maricopa County. "With images every 4 meters we are able to see the sides of structures and can gauge the width of overhangs and porches that other aerial images sometimes struggle to accurately measure."

"Maricopa has been able to recover over $12 million in new construction tax revenue on an investment of over a little over $1 million, [with] less staff time, in the desktop review process to date."

CHAPTER 4
Value Mapping and Geographical Analysis

1 | Introduction

The behavior of at-hand data and information drives decision making in virtually any environment. In the assessment community, the monitoring of data allows taxing jurisdictions to identify when and where adjustments are needed in order to produce accurate, fair estimates of value. Projecting data into a visual depiction, a process known as **data visualization**, allows assessment professionals to conceptualize trends and understand relationships that might be less obvious when the data are in their raw form, which in turn allows them to make more informed decisions. If data visualization is the lens for data analysis, geographic data visualization, or **geovisualization**, is the bifocal.

2 | Geovisualization

2.1 | Pre-Valuation Quality Assurance

Simply stated, data visualization helps to tell a more complete story. The inclusion of extreme or incorrect data (usually referred to as **outliers**)—whether the product of human error or other external phenomena—can result in subpar valuations that could prove quite costly for an assessor's office. In any assessment office, charts and graphs are used to show and examine the relationships between certain characteristics (e.g., size, condition, location) and value. Such visual inspection allows appraisers to identify potential outliers, which can then be corrected or removed to promote more accurate, uniform valuations.

Geovisualization adds another dimension in which relationships can be analyzed. By plotting property characteristics, valuations, and other data on a map, appraisers can gain previously untapped insight from analyzing variations by location. Outliers can be identified easily, adding another stage of quality control to the valuation process. Properties can then be inspected remotely to ensure that they are properly coded. *Figure 4.1* shows properties coded as "waterview" or "waterfront."

On this map, a blue dot represents a waterfront property, while a red dot indicates a property with a view of the water. An aberrant blue dot on a landlocked parcel or a red one on the water's edge would stick out, and could be identified almost immediately, allowing for correction and appropriate adjustments to be applied. Identifying miscoded properties without the aid of a map would be significantly more difficult and time-consuming, and might even go unnoticed.

Similarly, *Figure 4.2* shows properties plotted with their respective condition. By using geovisualization, appraisers who are familiar with certain areas can more easily identify potentially miscoded grades (such as a "very poor"-coded home located in a high-end community, or an "excellent"-coded property in a largely dilapidated part of town).

Appraisers can identify which homes have potentially incorrect or missing data and avoid the application of erroneous adjustments, which could result in an appeal of the final valuation.

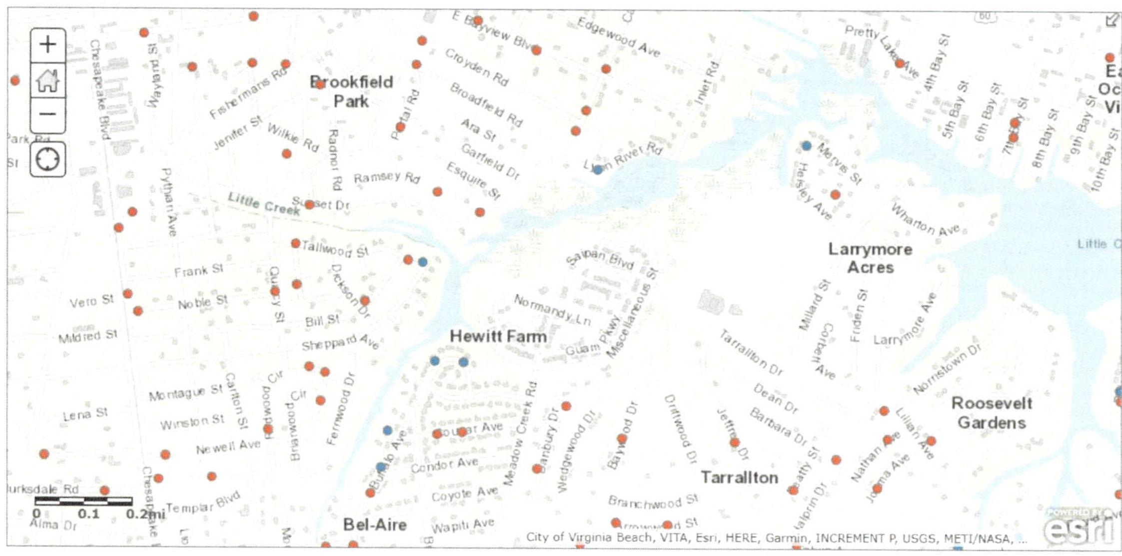

Figure 4.1: Map of Waterview vs. Waterfront Properties. *Source:* Esri and its licensors

Figure 4.2: Quality Grade. *Source:* Esri and its licensors; City of Fargo, Esri, DigitalGlobe, GeoEye, Earthstar Geographics, CNES/Airbus DS, USDA, USGS, AeroGRIND, IGN, and the GIS User Community

By remotely identifying areas that are in need of inspection or verification, **assessors are able to more efficiently allocate staff and resources associated with data collection,** saving both time and money.

Another useful feature of geovisualization is that multiple variables can be plotted simultaneously. *Figure 4.3* shows several house sales by both sale price (represented by size) and age (represented by color). Many different combinations of property data variables can be displayed in this manner.

Geovisualization gives jurisdictions the tools they need to better understand their markets and to identify potentially problematic areas, which ultimately promotes the accuracy of their valuations. The following gallery (containing *Figures 4.4* through *4.12*) offers a wide variety of ways that geovisualization could be used to equip an assessor with better information.

3-D GIS takes geovisualization a step further by visually conveying locational fluctuations. Any numeric value respective to a property (e.g., total living area, sale price, acreage) can be displayed as a sort of "elevation"—that is, with peaks and troughs representing comparatively high and low values, respectively. 3-D GIS is also used to demonstrate literal changes in height (such as office building stories) or elevation. With 3-D GIS, the relationships between viewsheds and value become much more apparent.

Please see Case Study 4.1 ("3D Modeling in GIS") at the end of this chapter for more information.

Figure 4.3: Map of Houses by Sale Price and Age.
Source: Esri and its licensors; City of Fargo, Esri

Figure 4.4: Foreclosure Count. *Source:* Esri and its licensors

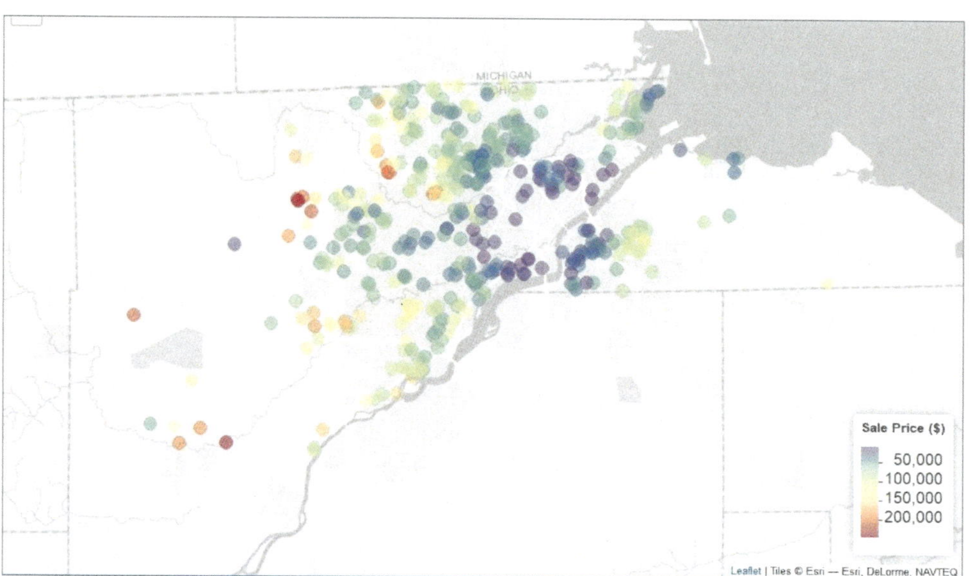

Figure 4.5: Price Map. *Source:* Esri and its licensors. Created in R Using the leafletR package. R package citation: Graul, Christian (2016): leafletR: Interactive Web-Maps Based on the Leaflet JavaScript Library. R package version 0.4-0, http://cran.r-project.org/package=leafletR.

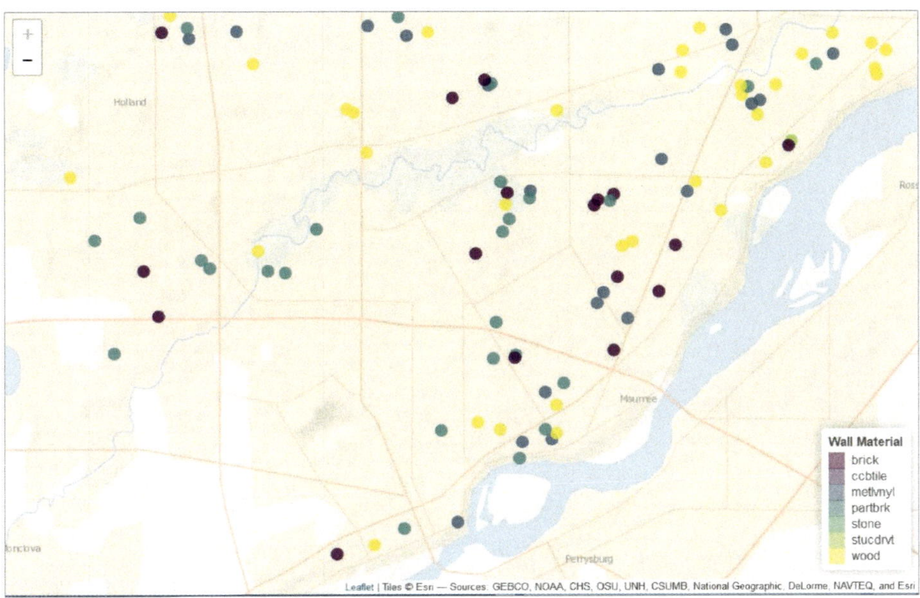

Figure 4.6: Wall Materials. *Source:* Esri and its licensors. Created in R Using the leafletR package. R package citation: Graul, Christian (2016): leafletR: Interactive Web-Maps Based on the Leaflet JavaScript Library. R package version 0.4-0, http://cran.r-project.org/package=leafletR.

Figure 4.7: Two-mile Buffer for Residential Comparable Sales Finder. *Source:* Esri and its licensors

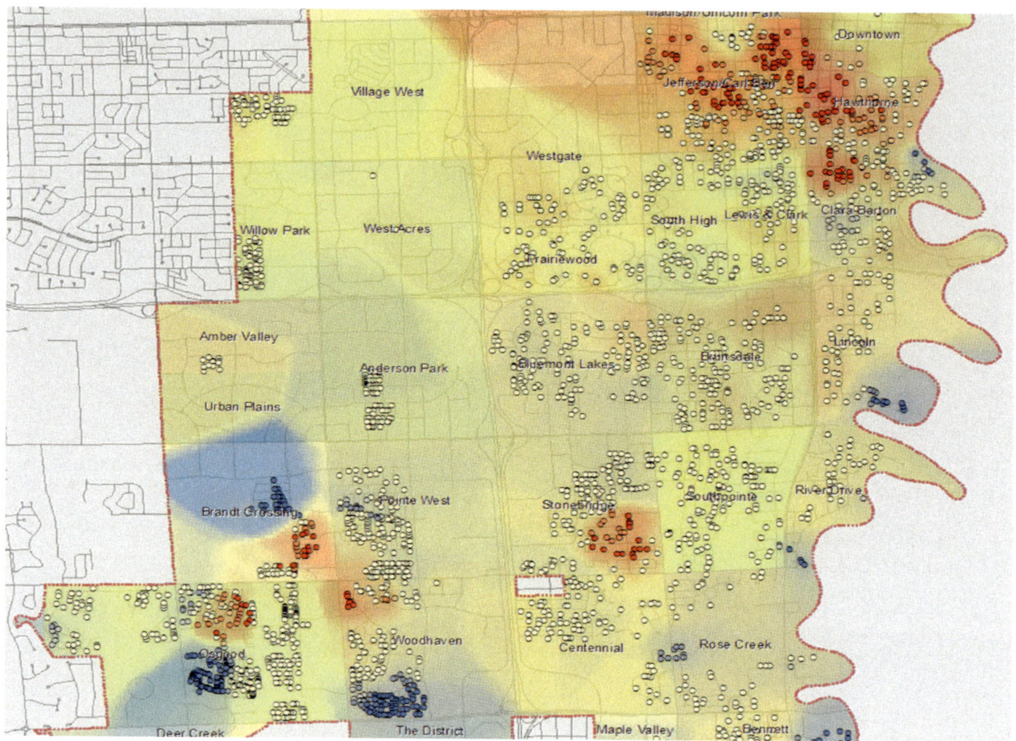

Figure 4.8: Clustering Sales Ratios. *Source:* Esri and its licensors; City of Fargo, Esri, Pro West & Associates

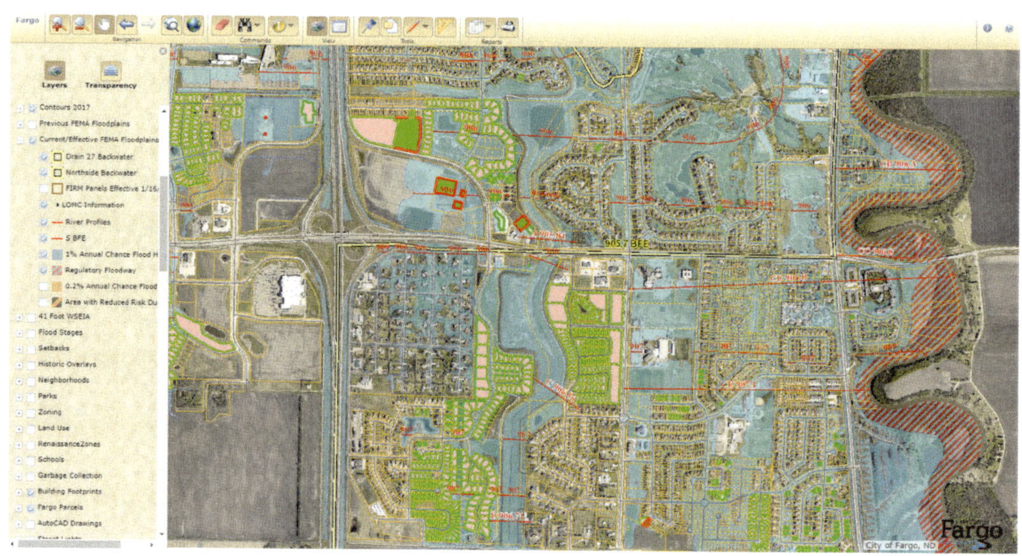

Figure 4.9: Public Info Floodplain Map

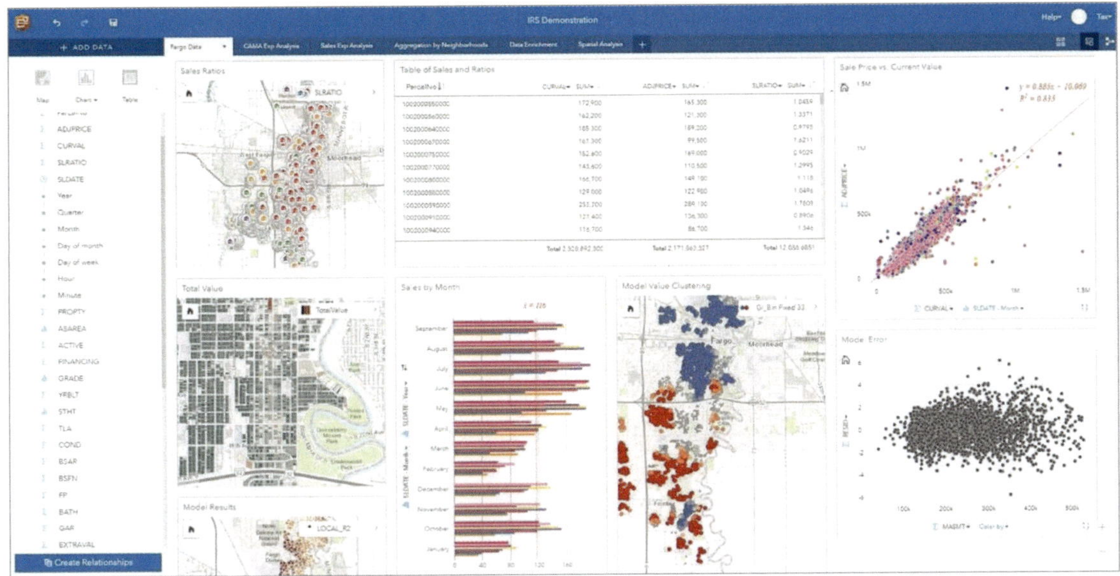

Figure 4.10: Esri Insights Dashboard - Property & Assessment Data. *Source:* Esri and its licensors; City of Fargo, Esri

Figure 4.11: Soil Type. *Source:* Esri and its licensors; Esri, DigitalGlobe, GeoEye, Earthstar Geographics, CNES/Airbus DS, USDA, USGS, AeroGRIND, IGN, and the GIS User Community.

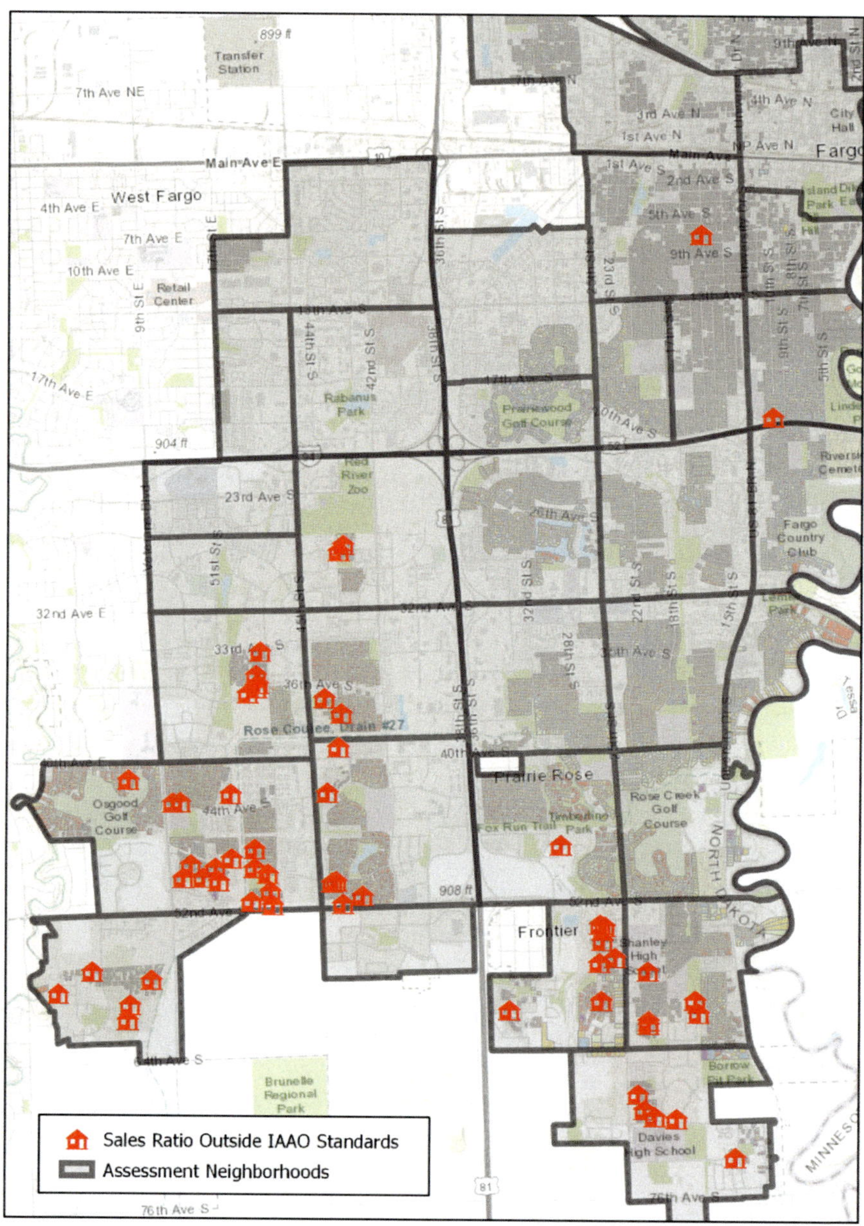

Figure 4.12: Unsatisfactory Assessment-to-Sale Price Ratios. *Source:* Esri and its licensors; City of Fargo, ND , Esri, HERE, Garmin, Intermap, increment P Corp, GEBCO, USGS, FAO, NPS, NCRAN, GeoBase, IGN, Kadaster NL, Ordnance Survey, Esri Japan, METI, Esri China (Hong Kong), swisstopo, OpenStreetMap contributors, and the GIS User Community.

REAL WORLD APPLICATION: Stearns County

Author: Randy Lahr, SAMA

Senior Appraiser, Stearns County Assessor's Office

In Stearns County, Minnesota, we have a very diverse land mix. As Stearns County is in the top 5 of agricultural producers in Minnesota, including dairy, poultry, swine, beef, and grain farming, land is put to many different uses likely because the soils vary so much throughout our large county (see *Figure 4.13*). With this being the case, it is vital that county assessors understand this and analyze sales to reflect this in their market values.

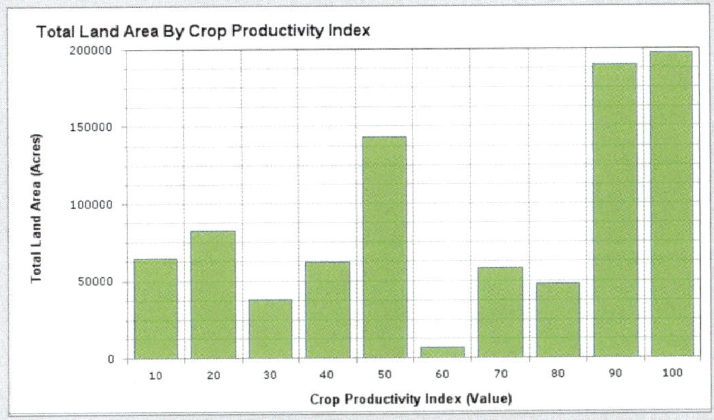

Figure 4.13: Total Land Area by Crop Productivity Index. *Source:* University of Minnesota (Minnesota Land Economics)

Up until recently Stearns County, along with many counties in the state, had been using a soil quality rating system called CER (crop equivalency rating). This is an old rating system and is no longer being supported. The University of Minnesota has implemented a new supported soil rating system called CPI (crop productivity index), which rates lands based on their ability to produce a bushel of corn, the predominant crop in Minnesota. Stearns County began using this new rating system in 2010 when we were integrating to a new internal GIS.

For years we had seen farm auction posters indicating that better CPI of the soils corresponded to higher prices. This was enough to convince us to improve our assessment by switching to the CPI rating. In the first year of implementation, it tightened up our sales assessment ratio by about 2%. For us, the first step was to decide where the CPI rating would fall within our current rating system. We were using an A1, A2, A3, B1, B2, B3, C1, C2, and C3 rating system so we wanted to bracket the CPIs within those categories. In our county we studied the sales of different soils and determined that once a soil rating got to below 50, typically sandy loam soils (light ground), there seemed to be little difference in sale price/acre. We established this as our lowest grade rating, C3 (see *Table 4.1*). From there we bracketed the ratings by price paid per acre of tillable land.

Table 4.1: Grade Ratings by CPI (Source: Stearns County Assessor's Office)

GRADE	CPI
A1	96 - 100
A2	90 - 95
A3	85 - 89
B1	80 - 84
B2	75 - 79
B3	70 - 74
C1	60 - 69
C2	50 - 59
C3	0 - 49

After establishing what the values should be per CPI rating, based on sales, the next step was to establish the CPI for each acre of tillable land in the county. This is where GIS was crucial. A requirement of CPI calculation is to clip the soil map layer to our field shape or parcel layer. The soil map layer can be downloaded for free by county at **http://www.mngeo.state.mn.us/chouse/soil.html#county**. By having a program in which we could draw out the field boundaries and then have it give us the soil types within that boundary or polygon, plus display the CPI rating and the weighted average CPI for that field or section of a field, made entering the data into our CAMA system a relatively easy transition.

Figure 4.14: Total Ac. by Weighted CPI. *Source:* WSB & Associates, Inc.

By having the CPI soils data joined to our parcel map layer it has allowed us to identify the quality of soil on each field and parcel within our county. This has tightened up our sales assessment ratio and promoted better equalization across all parcels in the county.

2.2 | Post-Valuation Quality Assurance and Defense

Diagnostics are used throughout a valuation cycle to identify optimal approaches to value and to promote accuracy, equity, and fairness of values. Striving to meet such principles is important for a number of reasons:

- Accurate, equitable, and fair valuation fosters public acceptance and trust, which is vital for any successful property tax regime, particularly in countries where ad valorem property tax is a new and potentially controversial concept.
- In countries whose property tax systems are more tenured, these principles help assessment offices to reduce costs associated with subpar valuations, such as the need for additional staff, legal fees, and court costs.
- For elected government officials, the integrity of their valuations could even have a direct impact on careers and re-electability.

IAAO maintains various standards that provide guidance and best practices with respect to property tax assessment and administration. The *Standard on Ratio Studies* outlines statistical tests that measure valuation accuracy, equity, and fairness. Many governments and courts throughout the world now hold local taxing authorities accountable to statutory requirements with respect to these tests.

These maps also help to identify potential locational effects on value that are perhaps not as apparent in tabular form. For example, it might be difficult to discern that properties near an airport have high **assessment-to-sale price** (**ASRs**), but when ASRs are displayed on a map, such patterns are more easily detected. An appraiser might also notice

When these statistics are plotted on a map, it helps assessors identify which values are "behaving" and which areas might need additional consideration, such as more or updated data and subsequent desktop review or "boots-on-the-ground" in the form of data collectors or appraisers.

that ratios are particularly high or particularly low in a given area and inspect those areas accordingly. Once physically present, the appraiser might notice subtle nuances in this area that are causing fluctuations in value, for example, blighted nearby properties, noise from a practicing high school marching band, particularly well-manicured and maintained landscaping, and so on. This information can then be used to create necessary adjustments to property values in similar areas, promoting accuracy, equity, and fairness. *Figure 4.15* demonstrates how "before and after" maps can convey an improvement in ASRs.

The **coefficient of dispersion** (**COD**) is a statistical test used to measure valuation uniformity—specifically, how precise ASRs of a particular group of sales are with respect to the median ASR. The lower the COD, the higher the uniformity, and vice versa. When CODs are calculated for submarkets (e.g., neighborhoods, boroughs), fluctuations in uniformity can be examined across an entire municipality. *Figure 4.16* shows how mapping COD can be used to evaluate uniformity by location.

Another concept related to the fairness of property valuations that is addressed in the *Standard on Ratio Studies* is vertical equity. Vertical equity suggests that properties, regardless of price, are valued equally—in other words, that ASRs of a given group of sales remain consistent across lower-, middle-, and upper-priced properties. If ASRs are not consistent across all price points, vertical inequity is present.

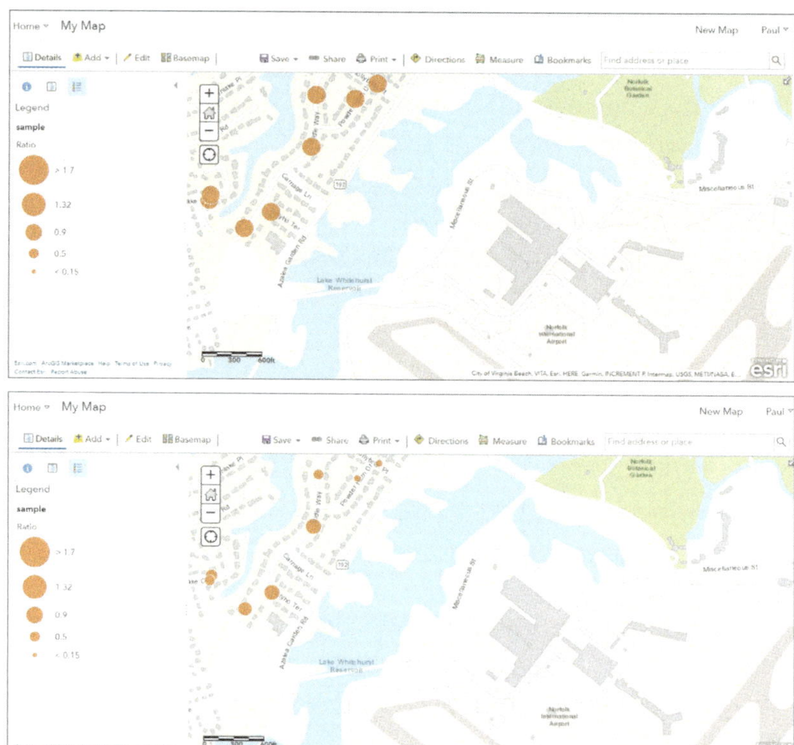

Figure 4.15: The above before and after maps demonstrate an improvement in ASRs after negative adjustments are made for airport proximity. *Source:* Esri and its licensors. Portrayed values are based on simulated, not actual, assessed values.

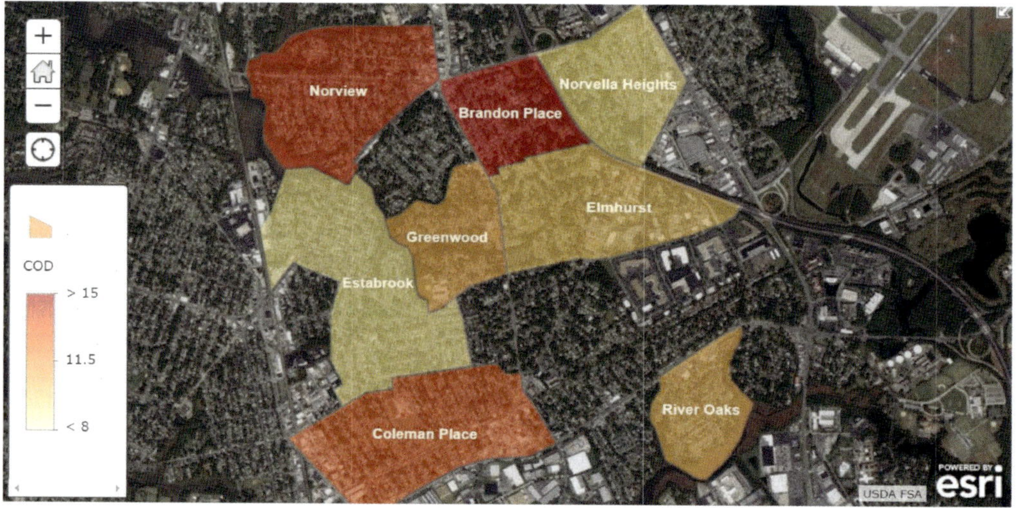

Figure 4.16: Mapping COD. *Source:* Esri and its licensors. Portrayed values are based on simulated, not actual, assessed values.

Figure 4.17: Mapping PRD. *Source:* Esri and its licensors. Portrayed values are based on simulated, not actual, assessed values.

There are a number of statistical techniques used to detect vertical inequity, and also to detect regressivity and progressivity, specifically. Such tests can be plotted on a map to identify areas with particularly serious offenses, as demonstrated in *Figure 4.17*.

By identifying the specific areas that need additional consideration and analysis, assessment offices are able to save time and money by only concentrating on areas whose valuations are not yet satisfactory. Plotting ratio studies and other statistics on a map also aids in the specification of mass appraisal models, helping to identify which models and variables are appropriate for each submarket.

The two types of vertical inequity with respect to property assessments are as follows:

Progressivity, where ASRs are comparatively lower for lower-priced properties

Regressivity, where ASRs are comparatively lower for higher-priced properties

Since ratio study maps help demonstrate the superiority of one valuation approach over another in a relatively clear manner, they are particularly useful for defending values. "Before and after" maps can be created to justify the application of a particular adjustment or valuation approach. Returning to our previous example of creating value adjustments for properties located near an airport, ratio study maps could be used to illustrate a bordering neighborhood's improvement in valuation uniformity after an adjustment for noise was made.

Maps can be used to demonstrate to a taxpayer or courtroom that the values for the neighborhood of a particular property are in line with industry standards outlined by IAAO, and are superior to other approaches to value that were considered. This reinforces your office's competence and builds trust with potentially leery parties.

REAL WORLD APPLICATION: Foreclosures

Authors: Paul Bidanset; Michael McCord, PhD; and Peadar Davis, PhD, MRICS

Foreclosures can be costly for assessors. There are a number of reasons why foreclosures might negatively impact the value of properties in the same market. Perhaps two of the more prominent reasons are: (1) reducing demand (with blight and potentially associated crime rates) and (2) increasing the stock of available homes, potentially even as cheaper alternatives or substitutes to non-foreclosure sales once they are released onto the market. Valuations can get trickier when foreclosures are perhaps "accidentally" chosen as comparable when determining sales prices or assessment values. As prices and values are pushed down, spillovers cause jurisdictions to forgo a larger tax base; Immergluck & Smith (2006) estimate that foreclosure spillover effects in Chicago reduced the overall tax base by approximately $598 million dollars from 1997 to 1998, with an average reduction of $159,000 per foreclosure.

A previous IAAO Academic Partnership Program paper evaluated how to properly account for foreclosures and sales-after-foreclosure in valuation models to improve valuation performance (e.g., uniformity, accuracy, equity). The data evaluated the market of Norfolk, Virginia, between the years of 2010 and 2012.

The authors created a model with variables that considered each sale's proximity to nearby foreclosed properties and sales-after-foreclosure, and the ratio of valid sales to foreclosures for each property's given neighborhood. Their model reduced the coefficient of dispersion (COD) for 41 out of the 85 neighborhoods with a maximum reduction of 22%.

However, while these foreclosure variables improved valuations for neighborhoods with a higher relative amount of foreclosures, they made valuations worse in the other neighborhoods (this because models can actually suffer from the inclusion of "irrelevant" – or variables that do not help explain what is impacting prices in the market) as was the case for these areas.

Figure 4.18 shows the neighborhoods that benefitted from this model, and how much each respective COD was reduced. Such maps can be used by assessors to identify how particular models or variables impact value performance for particular areas. This geovisualization can also be used to demonstrate why a particular adjustment was made to a taxpayer—e.g., why their neighborhood received an adjustment for foreclosures, and another neighborhood did not.

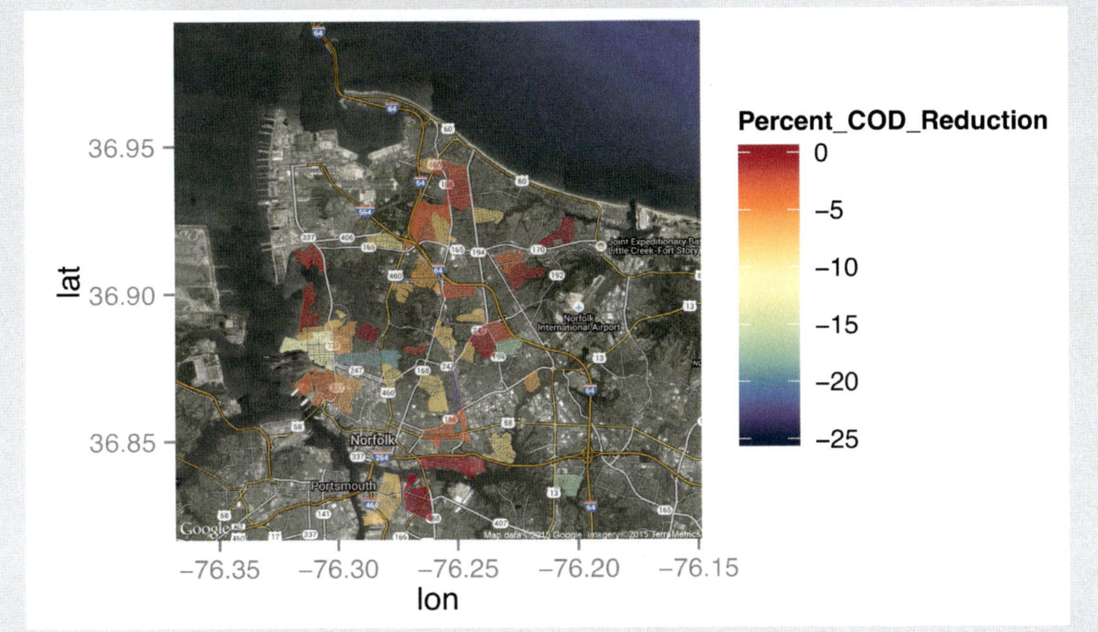

Figure 4.18: Percent COD Reduction. *Map data* © 2015; *Google Imagery* © 2015 TerraMetrics. Portrayed values are based on simulated, not actual, assessed values.

The original paper, titled *Spatially Accounting for Spillover Effects of Foreclosures in Automated Valuation Models to Promote Accuracy and Uniformity of Property Tax Assessments* by Paul Bidanset; Michael McCord, PhD; and Peadar Davis, PhD, MRICS was published in the Journal of Property Tax Assessment and Administration and can be accessed in the IAAO Research Library.

3 | Conclusion

Using geovisualization of assessment data as a tool during revaluations helps to ensure that valuations are accurate, fair, and defensible. When placed on a map, property characteristics help identify missing or incorrect data that might otherwise lie unnoticed in spreadsheets or other tabular formats—data issues that that can prove frustrating, costly, and time-consuming if left undetected.

Geovisualization of assessment data also allows assessors to identify areas of subpar valuations and effectively allocate resources only to the areas that require them, saving both time and money normally associated with data collection. When ratio study statistics are mapped, geovisualization enables assessors to optimize values by identifying necessary location-specific adjustments, and also increases the defensibility of valuations by providing clear, easy-to-understand visual aids. In summary, geovisualization promotes the integrity of valuations at all stages of the valuation process.

Case Study 4.1

3D Modeling in GIS

Author: Michael Lomax, Director • Assessment, Esri Canada

LIDAR

LiDAR (Light Detection and Ranging) is another advancement that is driven from and integrates with GIS. Users are now using this technology to determine building heights, HBU analysis, street-front measurements, and property characteristics like "roof shapes." Integration within geographic-assisted mass appraisal (GAMA) systems allow the user to automatically create various property attributes from the data and populate those data fields which enable more precise valuation calculations and more accurate valuations as a result.

Figure 4.19: 3D Model A

Figure 4.20: 3D Model B

SPATIAL ANALYSIS: SALES, NEIGHBORHOODS, MARKET TRENDS

Spatial analysis is further enhanced with 3D GIS techniques by providing the ability to view information in a 3D context which are physically correct to the strata plan and spatially correct to the GIS. In the below screen capture the user can see and thematically map the 3D scene by various property attributes or economic information like in this case sales. This provides the user the ability to spatially select sale comps versus the traditional way of looking at line reports and/or spreadsheets where no spatial context exists.

VIEW AND OTHER ADJUSTMENTS

Traditionally adjustments are prescribed by appraisers or assessors and the rationalization of those adjustments are by subjectivity, appraisal knowledge and/or educated opinions. GAMA/3D GIS are now allowing the appraisers to actually geo-process the views (obstructed and unobstructed) to determine scientifically what exact view a condo unit, commercial unit, or building has. This has led to less subjectivity, easier valuation defense and more accurate data as a result.

3D MODELS: PHYSICALLY AND SPATIALLY CORRECT

Database systems that incorporate the use of 3D technology is another technological evolution that is quickly being adopted by use of assessment professionals worldwide, by combining physically correct geometries (e.g. strata, condominium plans) with spatial accuracy (3D GIS x, y, & z coordinates) which results in a 3D GIS display of stacked titled properties. This technology allows strata/condo/multi-unit building plans to be converted to accurate 3D physical and spatially correct polygons where data are reviewed, analyzed, and validated where the properties can then be valued "en mass" all within the platform.

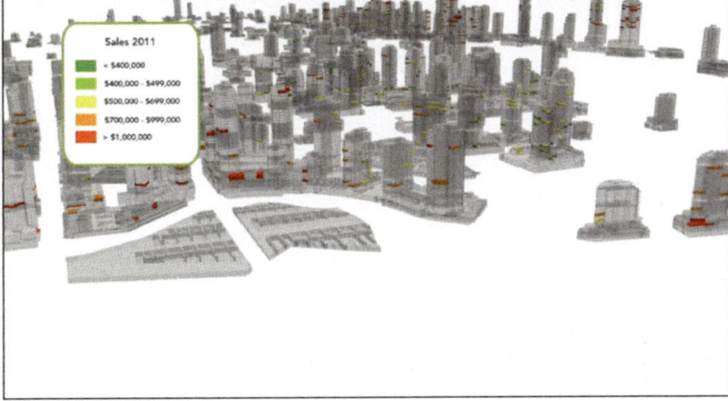

Figure 4.21 (top): 3D Map by Sales A
Figure 4.22 (bottom): 3D Map by Sales B

This type of new technology provides that same GAMA integration of building sketches, imagery, data, and analytics through said platform. It fully integrates GIS with CAMA, providing a single platform to perform a wide variety of activities without the need to transfer data between systems, subsystems or viewers. The result is improved data integrity, more accurate assessments and less time required to complete the requisite assessment and valuation work. Technological advances have thus allowed the capability to render complex 3D scenes which incorporate and leverage the power associated with complex 3D buildings and scenes. A complex building is where both the building and units are physically correct (correct building

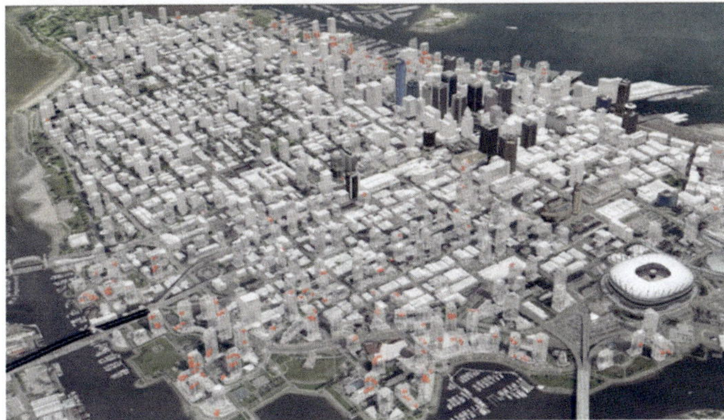

Figure 4.23 (top): Sample 3D Maps A
Figure 4.24 (bottom): Sample 3D Maps B

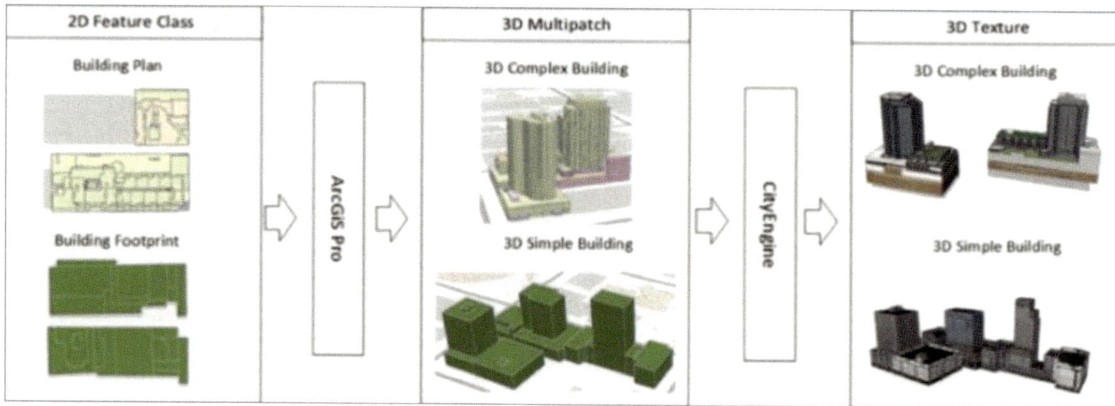

Figure 4.25: Simple vs Complex Buildings

Figure 4.26: 3D Rendered Buildings

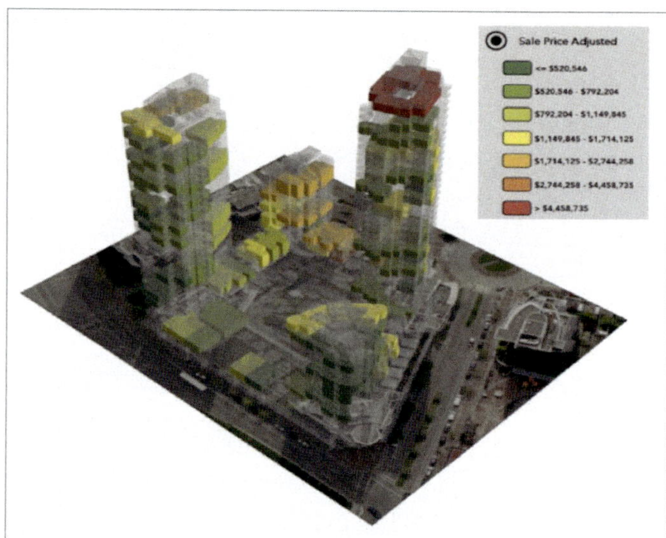

Figure 4.27: 3D Analysis

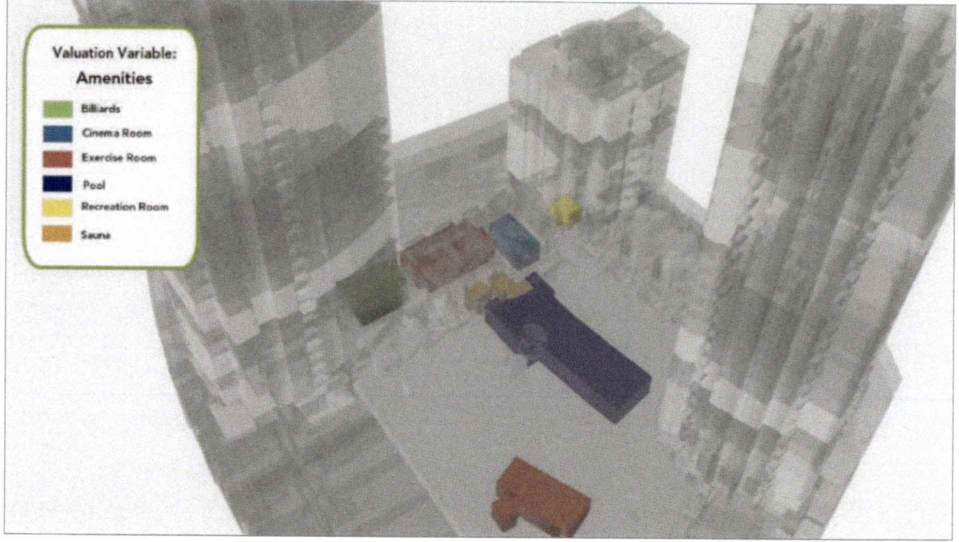

Figure 4.28: 3D Valuations & Comparable Sales Selections

and unit measurements) and spatially accurate (accurate point on the earth for each unit). These advances and uses of technology have moved the assessment industry from the use of "simple buildings" to the use of "complex buildings" not only for the assessment industry but also for the real estate industry as a whole (*Figure 4.22*).

The input data sets are converted to a 3D web scene that supports visualization of units and buildings. The 3D web scene provides different visualization, data and analytics at different extents. Data that can be viewed includes:

- Sale Prices
- Assessment Values
- Assessment to Sales Ratios (ASRs)
- Listing Prices
- Building Data
- Unit Data
- Economic Data
- Values
- Analytical Data (items such as floor adjustments, view adjustments, unit adjustments)
- Swipe between Appraiser's Office Assessment data and GIS "Geo-processed" data

COMPLEX 3D GIS BUILDINGS AND SCENES

Technological advancements and innovations have enhanced the capability to generate complex 3D building renditions moving from simple buildings (think Google Earth) to more complex buildings rendered from the strata plan and/or condominium declaration (*Figure 4.26*).

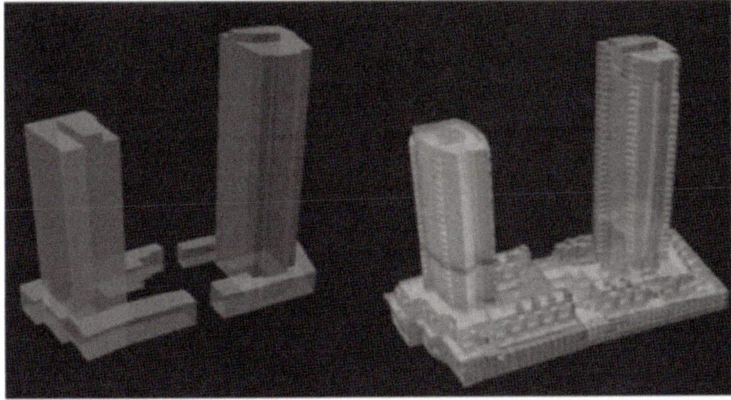

Figure 4.29: Simple Building (Left) and Complex Building (Right) 3D Models

City governments are now obtaining building plans in electronic formats and as a result they are easily consumed in a 3D GIS or GAMA system (digital to digital file conversion). For historical plans there are methodologies for converting paper residential or commercial plans (condo/strata/leasehold/rental) into complex 3D models (*Figure 4.27*). Source data can either be paper (PDF) or digital (CAD) or building strata plans. BIM model of the architecture plans can also be used as well for development of3D complex models (*Figure 4.28*).

Figure 4.30: File Conversion—Paper (Strata/Condo Plan) to Digital (3D GIS)

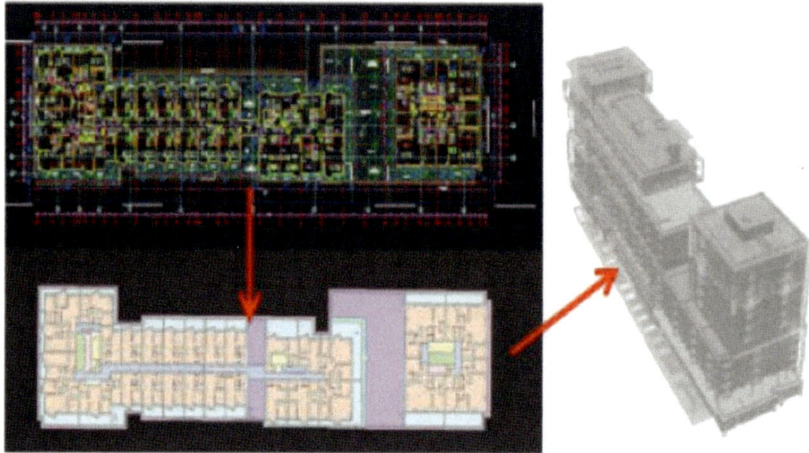

Figure 4.31: File Conversion—Digital (CAD/BIM Models to 3D GIS)

3D GIS DISPLAY

As illustrated, change is quickly occurring in the technologies needed and used by the Assessment Industry. Greater demands for visualization, on demand analytics and sophisticated map-based information are becoming expected by the appraisers as well as the public when they access assessment data. The new current reality of assessment technology incorporates a greater and more integrated approach to the use of GIS for the assessor. GIS technologies have advanced dramatically and continue to advance as the leading source of location based (spatial) display and analytics. The evolution of CAMA over the past 10 years has seen the assessment industry placing a greater reliance on GIS in property database administration. Consequently, the use and dependence on GIS has grown dramatically within the assessment industry.

GIS adds value to CAMA systems in both 2D & 3D, such as an appraisal model, which can place added value on property that has, for example, frontage on a golf course or lake. If the parcel data and land-use data are maintained in a geodatabase, the frontage calculation is simple and easy using the spatial intelligence of the GIS. Integrating GIS and computer-assisted mass appraisal

(CAMA) enables the tax assessment function to be concurrent with spatial data that is relevant to the tax valuation model. It also supports the creation and maintenance of a more accurate land records basemap using the tools and functions of GIS and provides a single repository of parcel geometry and descriptive data supporting workflow, updates, and mass appraisal input.

The following screen shots of "assessment analyst, aka: Desktop Review" (*Figure 4.29*) provide an illustration of how GIS is used as an integration platform tying specific assessment data and imagery together for the user.

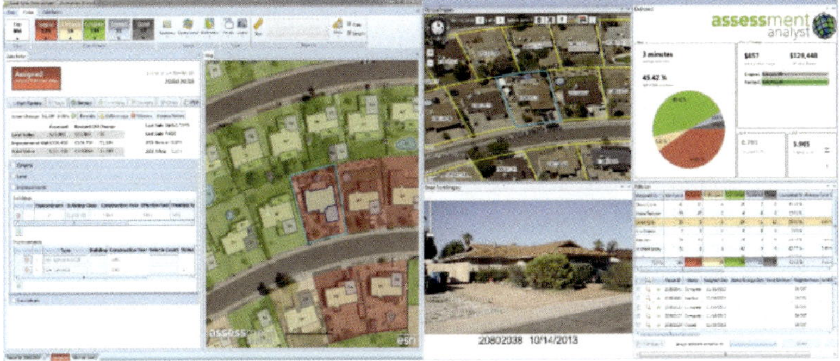

Figure 4.32: Screenshot I – Assessment Analyst, aka: Desktop Review

Figure 4.33: Screenshot II – Assessment Analyst, aka: Desktop Review

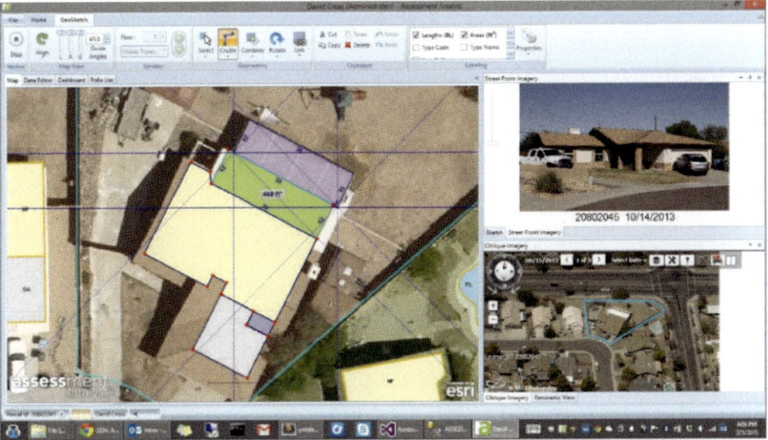

Figure 4.34: Screenshot III – Assessment Analyst, aka: Desktop Review

Case Study 4.2

Using Scripting/Programming Languages to Enhance Sales Ratio Analysis

Author: James Thomas, Assessor • City of South Portland, Maine

OUTLINE

This article suggests the use of scripting/programming languages, such as Python, to enhance the tools available to the assessor in performing sales ratio analysis. This article is not about using GIS spatial tools, but will use GIS to explain the Python tools developed as a framework for understanding how scripting languages can rapidly produce sales ratio data, based both on spatial selection and selections based on attribute queries. An understanding of spatial selection tools and linking (relating) tables in GIS software is assumed.

Sales ratio tools in the CAMA software that municipal assessors use are really quite good. Usually an assessor can create cross-sections and stratifications of sales data by setting filters or using SQL statements fairly easily. What is absent from these functions is a way to use "space" as a method to generate sales ratio data. Following is a discussion of three scenarios where a python scripting tool (the Tool) can quickly calculate ratios and CODs for areas of interest.

THE PYTHON TOOL AND DATASET

The dataset used consists of two objects: 1) a parcel feature class attributed with property characteristics, such a valuation neighborhood codes, site adjustments, building styles and other property characteristics imported from a CAMA system, and 2) a table of time adjusted residential property sales data (Salesfile) with similar attributes as above, including attributes representing Assessments and Sales Prices. The use of a GIS linking feature to relate the parcel data with the Salesfile solves the problem of "One to Many" sales per parcel. (i.e. permits the use of time adjusted paired sales.)

As an analyst makes a selection of parcels based on attribute data or spatial selection, the tool asks the analyst for a "Summary" field. (neighborhood, building style, grade, etc) This entry is used to subtotal ratio data for each instance of the chosen field. For example, if the assessor chooses Grade as the summary field and the assessor selects a wide area of a jurisdiction, the Tool will calculate ratio data and a COD for data for each instance of the Grade within the selection.

SCENARIO I

The section of a jurisdiction shown on *Figure 1* shows a layer of 100 foot wide buffers (Band) covering a valuation neighborhood that is known to be growing in value faster the average area in the community. Site adjustments are currently used to adjust for a view the ocean. However from

Figure 4.35

Table 4.2

BAND	RATIO	QR	COUNT
3	72.21	15.41	7
4	75.44	12.31	12
5	74.51	6.72	3
6	73.19	16.13	7
7	77.46	16.41	10
8	84.49	13.28	13
9	74.12	12.62	14
10	77.15	4.18	5
Overall Ratio	76.66	13.55	71

sales surveys, the City suspects that while view sheds are significant value factors, mere distance (walking distance) to the shore is probably also a valid property factor.

Using the buffers to populate the Salesfile with Band values, the Tool generates the Table in *Figure 1*. As expected, the overall ratio is lower than the jurisdiction wide ratio of 83%, however it shows variation on the ratios between the bands, indicating that further study is warranted.

SCENARIO II

Developing adjustments for corridors between high value neighborhoods and lower value neighborhoods can be time consuming. *Figure 2* shows a red graphic between a higher value area, "CG" and a lower value area, "TH." In both neighborhoods the average sales ratio is about 87%. Using the graphic to select properties in the corridor and executing the Tool shows variation in the ratios of sales in the corridor, indicating the possibility of a slight downward "tug" in value of those properties in CG that border TH.

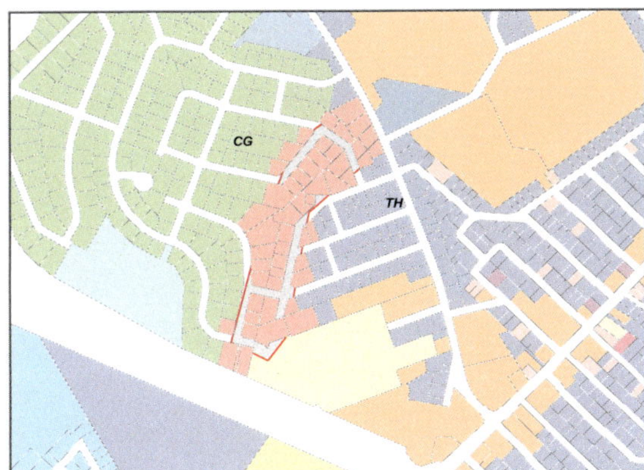

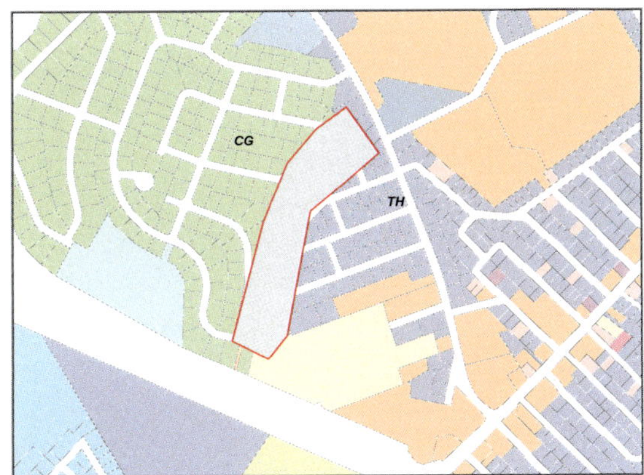

Figure 4.36

Table 4.3

S_NHBD	RATIO	QR	COUNT
CG	90.35	7.73	4
TH	86.95	16.89	3
Overall Ratio	88.89	11.88	7

SCENARIO III

Consider the image in *Figure 3*. The oval, the line and the circle represent random spatial selections, where parcels underneath the graphics could represent any type of property, residential, commercial, vacant. Consider that such selections can be done rapidly, literally the click of a button, and can be further filtered by property features such as SLA or any other characteristic found in the dataset. The table shows the ratios categorized by Site Area adjustments and show a 3% difference in the ratios, perhaps indicating further adjustments may be appropriate in one or the other site area.

S_NF_SI	RATIO	QR	COUNT
1_6	89.37	8.11	10
1_7	86.63	8.71	24
Overall Ratio	87.44	8.77	34

Figure 4.37

SUMMARY

Finally, in general, the Tool can be used to measure the overall performance of the total body of assessments with respect to statutory and best practice assessment standards. (i.e. uniformity in ratios between and within all classes of property.) In theory, if the body of assessments are completely uniform, the ratios would be equal wherever the analyst make a selection. Importantly, when they are not, the Tool plainly shows that as well.

CHAPTER 5
GIS and Mass Appraisal Models

1 | Introduction

An appraiser would be hard-pressed to find a real estate market that behaves the same way across any given location. The location of a property will have a large impact on the desirability of that property. Apartment units in a trendy, revitalized part of town that offer a "live, work, play" lifestyle, for example, will likely be able to charge higher rents than more utilitarian apartments situated on the outskirts of town. Availability, too, will vary by location. Prestigious, historic homes usually command a premium, but are also typically restricted to the areas of town that have withstood urban transformations and redevelopment. *Figure 5.1* provides a visual representation of how homes' ages can vary across a jurisdiction—one of many reasons why values will fluctuate by location.

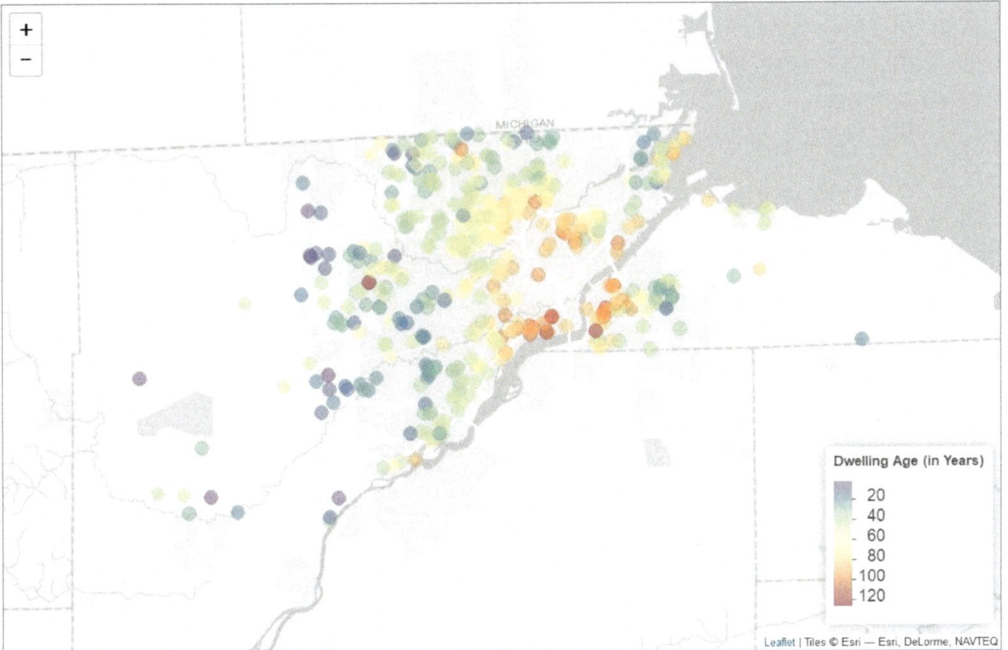

Figure 5.1: Age Map. *Source:* Esri and its licensors. Created in R Using the leafletR package. R package citation: Graul, Christian (2016): leafletR: Interactive Web-Maps Based on the Leaflet JavaScript Library. R package version 0.4-0, http://cran.r-project.org/package=leafletR.

Price fluctuations due to location might be experienced at a more granular level as well. For example, a home that sits on a busy intersection might not sell for as much as an identical home that's located on a safer, quieter lot just one mile down the road. Also, while one part of town depreciates, another might remain unchanged—or even appreciate—all for reasons created by, or related to, location.

This chapter will address the proper ways to improve real estate valuations, by accounting for location, through the use of GIS and spatial modeling.

2 | Location and the Market

2.1 | Supply and Demand

The degree to which a property's price might fluctuate based on its location can be boiled down to the two primary forces at play in every marketplace: supply and demand.

Supply refers to the availability of real estate stock—namely, the quantity of properties on the market, their characteristics, and the financial expectations of sellers and leasers. **Demand** refers to the desirability of a property in conjunction with the willingness and financial ability of buyers or leasers to pay for that property. The supply of and demand for a specific type of property will vary by location, and are influenced by the following factors:

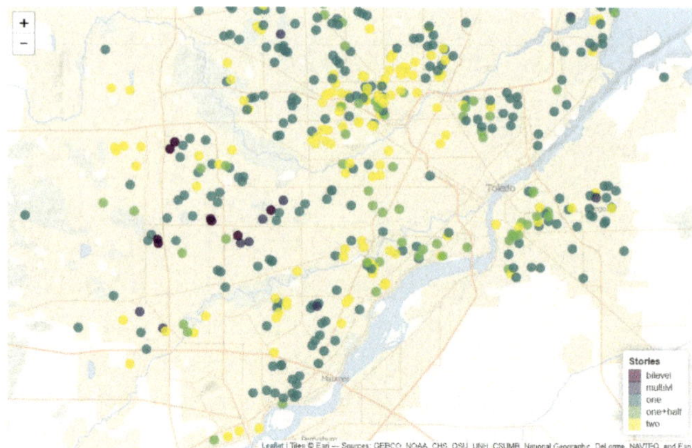

Figure 5.2: Property Characteristics Map. *Source:* Esri and its licensors. Created in R Using the leafletR package. R package citation: Graul, Christian (2016): leafletR: Interactive Web-Maps Based on the Leaflet JavaScript Library. R package version 0.4-0, http://cran.r-project.org/package=leafletR.

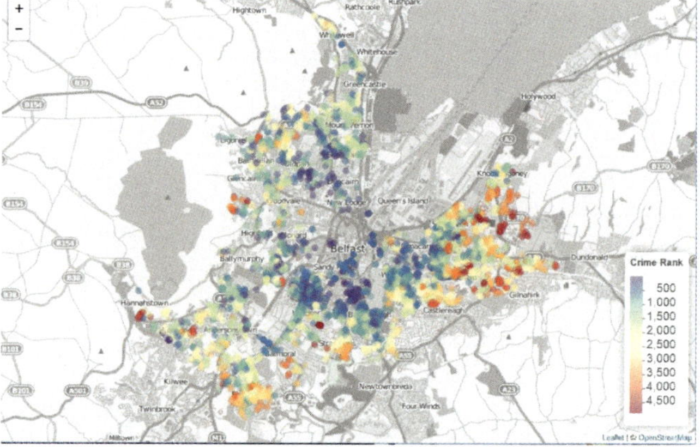

Figure 5.3: Belfast Crime Rates Map. *Source:* https://openstreetmap.org/copyright Created in R Using the leafletR package. R package citation: Graul, Christian (2016): leafletR: Interactive Web-Maps Based on the Leaflet JavaScript Library. R package version 0.4-0, http://cran.r-project.org/package=leafletR.

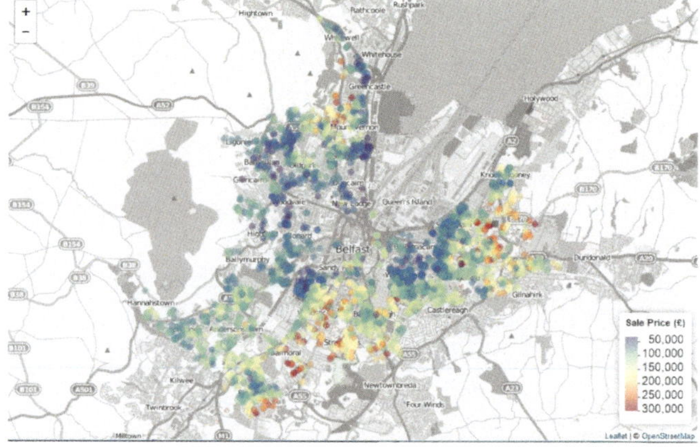

Figure 5.4: Belfast Property Values Map. *Source:* https://openstreetmap.org/copyright Created in R Using the leafletR package. R package citation: Graul, Christian (2016): leafletR: Interactive Web-Maps Based on the Leaflet JavaScript Library. R package version 0.4-0, http://cran.r-project.org/package=leafletR.

- Property characteristics (e.g., architectural style, size, view). *Figure 5.2* shows what a map might look like that organizes parcels by a specific property characteristic.

- Surrounding areas (e.g., school districts, proximity to shopping, crime rates). Taken together, *Figures 5.3* and *5.4* show how an area's property values are affected by the level of crime present in surrounding areas.

- The economic makeup of the buyers or renters associated with a location. Buyers and renters of relatively lower income, while they might possess a willingness to, are not likely to have the resources necessary to impact demand for (and therefore the price of) higher-priced properties. Similarly, the economic makeup of sellers and lessors in a given location will ultimately affect the asking prices and the accepted prices of property.

Supply and demand will vary across a jurisdiction and are affected largely by the local population's tastes, preferences, and what they can afford. Take data from any census or household survey, plot them on a map, and the patterns of income and population characteristics will become apparent. *Figures 5.5* and *5.6* illustrate the correlation between an area's income and its property values—one of many variables that result in locational differences in value.

As we discussed before, the desirability of a property's location will have impacts—either positive or negative—on its value. *Table 5.1* lists several examples of location characteristics across various property types, along with the theoretical impact of each characteristic on that property's value.

Before GIS can be incorporated into modeling, locations must be geocoded. GEOCODING is the process of assigning grid coordinates (e.g., latitude and longitude) to a location. Typically, this location is in the form of a street address. However, street names, postal codes, cities, and more can also be geocoded. Any point that is to be included in spatial modeling or analysis must be geocoded, including properties to be valued and any locations for which a proximity adjustment is to be calculated (e.g., distance to grocery store).

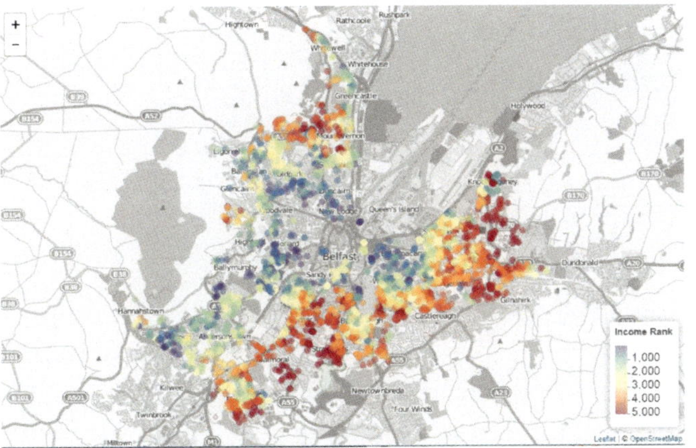

Figure 5.5: Belfast Income Rank Map. *Source:* https://openstreetmap.org/copyright Created in R Using the leafletR package. R package citation: Graul, Christian (2016): leafletR: Interactive Web-Maps Based on the Leaflet JavaScript Library. R package version 0.4-0, http://cran.r-project.org/package=leafletR.

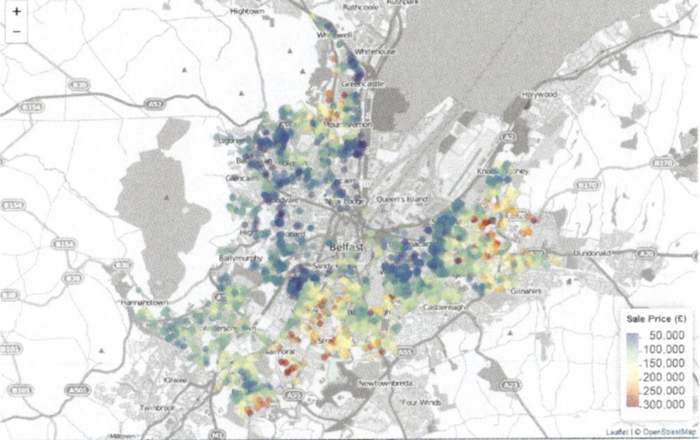

Figure 5.6: Belfast Sale Price Map. *Source:* https://openstreetmap.org/copyright Created in R Using the leafletR package. R package citation: Graul, Christian (2016): leafletR: Interactive Web-Maps Based on the Leaflet JavaScript Library. R package version 0.4-0, http://cran.r-project.org/package=leafletR.

Table 5.1: Impact of Location Characteristics by Property Type

PROPERTY TYPE	LOCATION CHARACTERISTIC	EXPECTED IMPACT ON VALUE
Single-family Dwellings	Well-performing school districts	+
	Proximity to golf courses and other recreational areas	+
	Proximity to shopping (e.g., grocers, retail)	+
	High noise pollution	-
	Proximity to landfill	-
	Higher crime densities or incidences	-
Apartments	Proximity to universities	+
	Proximity to restaurants and nightlife	+
	Proximity to parks and urban greenspace	+
	Low population and employment density	-
Retail	Higher densities of competitors	-
	Lack of accessibility or parking	-
	Proximity to public transportation	+
	Areas of higher consumer spending power	+
Industrial	Proximity to noise ordinances	-
	Lack of physical expansion opportunities	-
	Proximity to skilled workforce density	+
	Proximity to railways, ports, highways	+

This inherent variation of supply and demand across markets can result in wide-ranging fluctuations in value, presenting a number of hurdles that appraisers must overcome in order to accurately value properties. Fortunately, GIS provides appraisers with the tools necessary to quantify appropriate locational adjustments, promoting accurate and equitable valuations. These tools include spatial variables and spatial modeling methodologies.

2.2 | Regression with Spatial Variables

Spatial variables are used in regression-based automated valuation models (AVMs) (also known as computer-assisted mass appraisal (CAMA) models) to increase the models' ability to provide an accurate estimate of value. These variables are used to account for the large impact that location has on value.

2.2.1 | Locational Dummy Variables

In regression analysis, dummy variables (also known as "binary" or "indicator" variables) take on a value of either 0 or 1. When used in AVMs, they indicate whether or not a property has a particular attribute, with a value of 0 meaning "no" or "false," and a value of 1 meaning "yes" or "true." Variables may then be created to denote a property's location within a jurisdiction—for example, a neighborhood or school zone. They also denote location on a more granular level, such as whether or not a property is situated on a corner lot or cul-de-sac. The coefficient produced by the AVM for each of these locational dummy variables is then simply a location adjustment.

A good statistical "rule of thumb" is that for each dummy variable used, there should be at least five (but ideally 15) observations that belong to the classification. For example, if you wish to include a dummy variable to denote a waterfront property, there should be at least five waterfront properties in your data set. Many jurisdictions do not have enough sales to include a dummy variable for each neighborhood, and should therefore explore other modeling techniques to account for location.

2.2.2 | Distance Variables

When used in AVMs, distance variables can help quantify potential adjustments in order to account for a property's proximity to other locations, amenities, or phenomena. Positive adjustments might be needed for properties nearer to public transportation hubs, beaches, or desirable shopping areas, while negative adjustments might be needed for properties closer to higher crime density, foreclosures, or landfill. To quantify adjustments based on distance, variables that represent proximity are included in the model. The most common distance variables include:

- **Straight-line (Euclidean) distance variables** represent the distance between two points "as the crow flies." Their use is appropriate when the impact on value is not likely to be assuaged by a physical barrier, such as noise from a nearby airport or smell from a landfill. Straight-line distance variables are measured in meters, miles, kilometers, and so on.

- **Travel distance variables** represent the distance necessary to drive, walk, bike, etc., from one point to another. An example might be the number of kilometers there are between a house and a grain depot in a rural, agricultural area. Like straight-line variables, travel distance variables are measured in meters, miles, kilometers, and so on. They are appropriate to use when the locational impact on value is based on accessibility via a specific route where impedances such as stoplights are either nonexistent or minimal.

- **Travel time variables** represent the time it takes to travel between two points by driving, walking, biking, etc., and are measured in minutes or hours. An example would be the number of minutes it takes to walk from a restaurant to the nearest public transport. These variables should be used when locational impact on value is based on accessibility, specifically via a route (e.g., a road network) where impedances such as stoplights are abundant.

Figure 5.7 shows the differences in these three distance metrics.

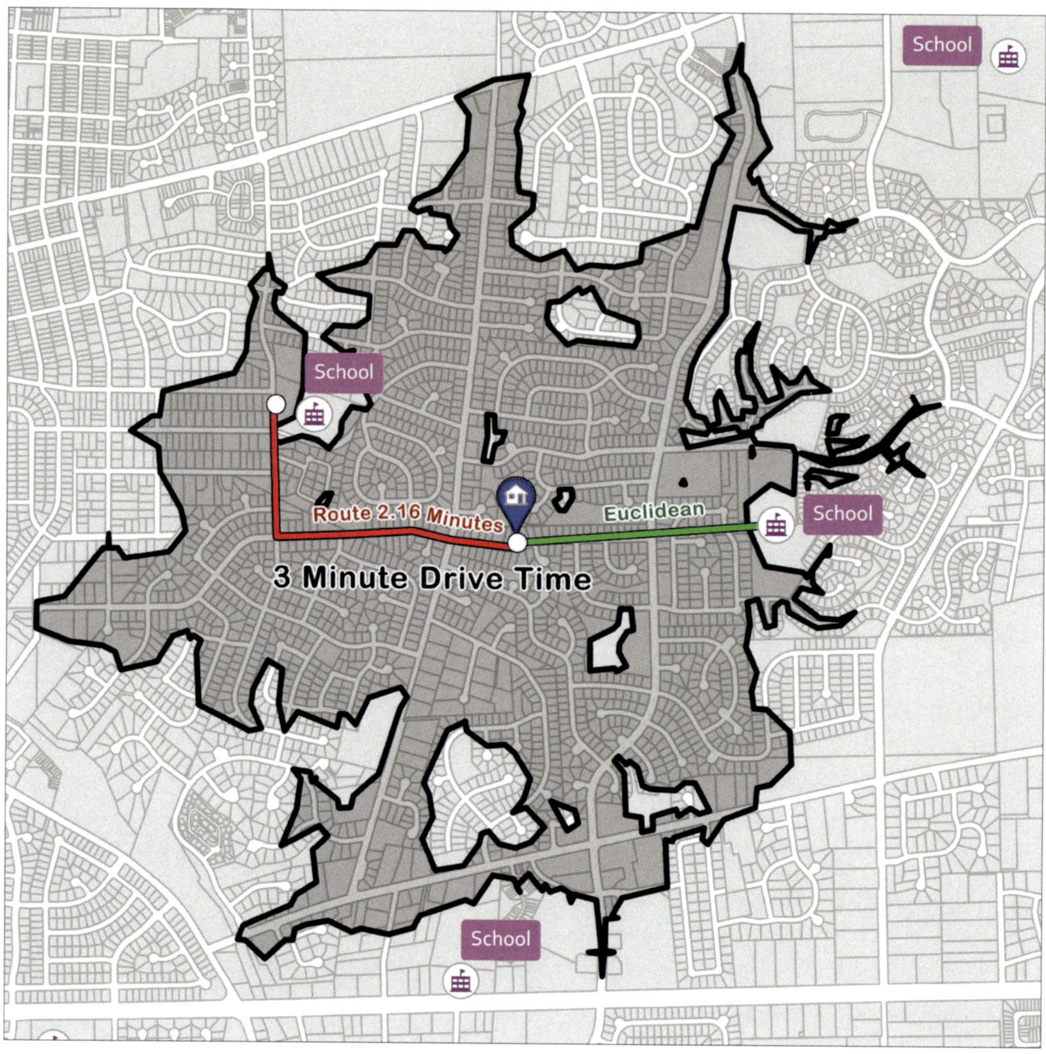

Figure 5.7: Map with Three Distance Metrics. *Source:* Esri and its licensors

The final distance variables that should be discussed, GIS buffer zones, encapsulate the specific distance around a location based on distance or time. Buffer zones can be created based on a fixed-distance diameter, in which case the buffer would be circular, or on travel distance/time, in which case the buffer would not be symmetrical. Once constructed, AVM variables can then be calculated. Some examples include "number of home break-ins within a 1-mile radius" and "whether or not there is a coffee shop within a 10-minute walking distance."

▸ 2.2.3 | Population Variables

As previously mentioned, tastes and preferences for real estate will vary depending on a location's population. For example, areas with younger populations might experience premiums associated with access to public transportation and nightlife, good school districts, or higher bedroom counts (for young families). Areas with aging populations, on the other hand, might experience higher premiums associated with access to golf courses and other recreational amenities, healthcare services, or single-story access (as they could have trouble climbing stairs now, or anticipate stairs becoming more difficult to navigate in the future). However, since demand is determined not just by willingness, but also by a financial *ability* to pay, prices will fluctuate across income levels.

Incorporating population data into AVMs can yield more accurate estimates of value by helping to account for the tastes, preferences, and economic ability of certain submarkets. Governments often publish free, open data, collected by a census and placed in geodatabases and shapefiles, that are ready to be utilized by GIS software. That said, the frequency with which these data are updated and the level of detail present will vary by country. *Figures* 5.8 and 5.9 provide examples of how an open data portal interface might look.

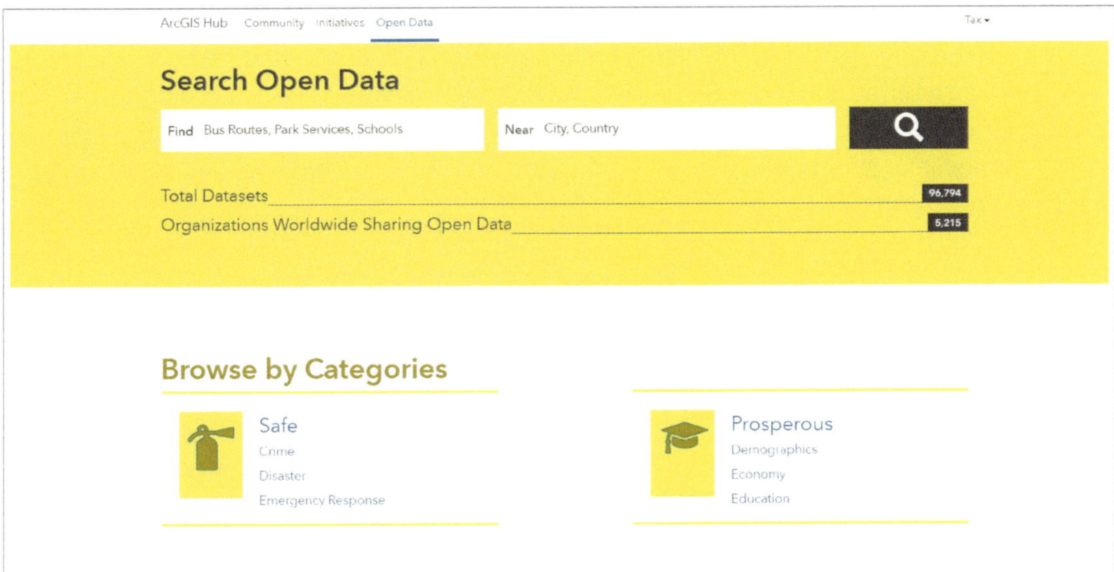

Figure 5.8: Esri's Open Data Portal ArcGIS Hub Example I. *Source:* Esri and its licensors

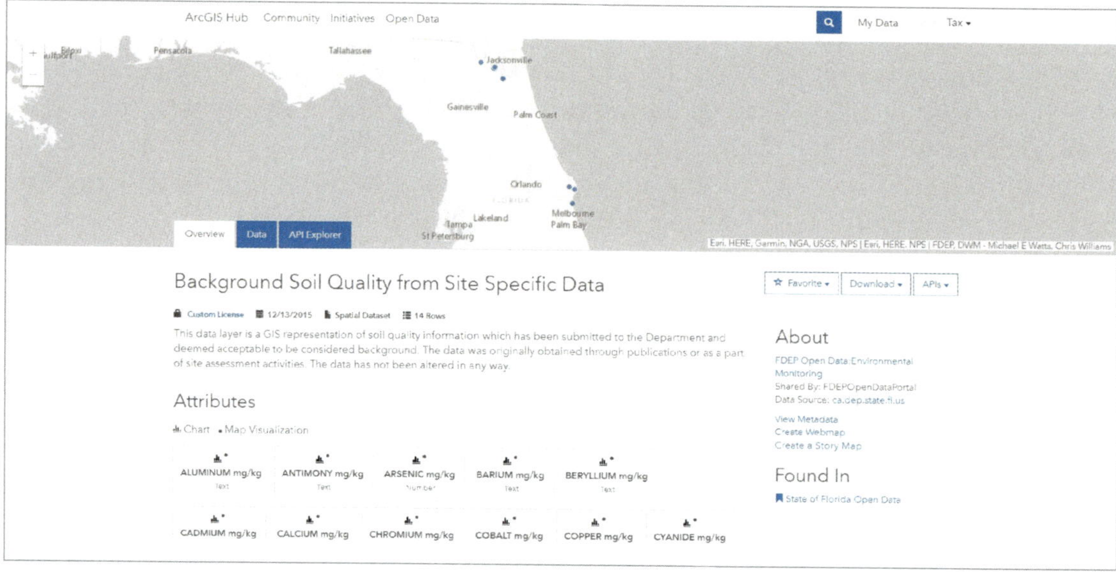

Figure 5.9: Example of Open Soil Data on ArcGIS Hub Example II. *Source:* Esri and its licensors

GIS software firms collect these and other data on a more regular basis and make it easy to access and append to spreadsheets, and even integrate with CAMA systems. Coefficients of population data variables in AVMs can be used as adjustments for various levels of population, education, and more. Inclusion of such variables in AVMs also acts as a quality assurance measure against over-valuing areas associated with low income.

Please see Case Study 5.1 ("Examining the Spatial Relationship Between Environmental Health Factors and House Prices: NO_2 Problem?") at the end of this chapter for a demonstration of a practical application of spatial modeling.

3 | Spatial Modeling Methodologies

More advanced spatial regression approaches are capable of modeling locational variations in value without it being necessary to include spatial variables. These methodologies use GIS coordinates (e.g., latitude and longitude) to predict values across a geographic plane, and include geographically weighted regression, spatial interpolation models, and spatial regression models.

3.1 | Geographically Weighted Regression

Geographically weighted regression (GWR) is a modeling approach that allows sales in closer proximity to a subject property to have more of an influence on value estimation than properties further away. Unlike an ordinary least squares (OLS) AVM that produces one regression output of diagnostics and coefficients, GWR produces one output for each sale location, resulting in model coefficients that can vary by location. GWR has demonstrated, in many instances, an ability to produce more equitable and uniform valuations than OLS-based AVMs. *Figures 5.10* and *5.11* demonstrate how these outputs compare to one another—*Figure 5.10* shows a sample GWR output, while *Figure 5.11* shows a sample OLS output.

```
****************Summary of GWR coefficient estimates:******************
                     Min.      1st Qu.      Median     3rd Qu.     Max.
   Intercept       -5.2489406  6.8654282  9.3995692 12.3330120 36.0760
   log.LandArea.  -0.0197078  0.1577087  0.1995856  0.2444668  0.5193
   log.Age.       -0.6557932 -0.1794831 -0.1356799 -0.0836074  0.2364
   Beds           -0.0941694  0.0277308  0.0560853  0.0840223  0.2414
   Baths          -0.0872653  0.0588050  0.1035204  0.1609858  0.3676
   log.Pop.       -1.3944241 -0.1102145  0.0406913  0.2019729  1.1618
   log.MedHHinc.  -1.8864048 -0.1489643  0.0663648  0.2748824  1.0729
   log.Grad_edu.  -0.9491666 -0.1829544 -0.0308542  0.1053319  1.1627
   RM6            -0.0181922  0.0030141  0.0071194  0.0119629  0.0342
   ************************Diagnostic information*************************
   Number of data points: 3232
   Effective number of parameters (2trace(S) - trace(S'S)): 733.3666
   R-square value:  0.87312
   Adjusted R-square value:  0.8358649
```

Figure 5.10: Sample GWR Output. Created in R using the GWmodel package. Citation: Isabella Gollini, Binbin Lu, Martin Charlton, Christopher Brunsdon, Paul Harris (2015). GWmodel: An R Package for Exploring Spatial Heterogeneity Using Geographically Weighted Models. Journal of Statistical Software, 63(17), 1-50. URL http://www.jstatsoft.org/v63/i17/.

```
Coefficients:
                Estimate Std. Error t value Pr(>|t|)
(Intercept)    11.229202   0.297486  37.747  < 2e-16 ***
log(LandArea)   0.202510   0.004479  45.213  < 2e-16 ***
log(Age)       -0.096402   0.006047 -15.943  < 2e-16 ***
Beds            0.051001   0.007323   6.964 3.99e-12 ***
Baths           0.200102   0.007708  25.960  < 2e-16 ***
log(Pop)       -0.075247   0.015517  -4.849 1.30e-06 ***
log(MedHHinc)  -0.057315   0.024182  -2.370   0.0178 *
log(Grad_edu)   0.015141   0.013808   1.097   0.2729
RM6             0.006706   0.001474   4.550 5.57e-06 ***
---
Signif. codes:  0 '***' 0.001 '**' 0.01 '*' 0.05 '.' 0.1 ' ' 1

Residual standard error: 0.2358 on 3223 degrees of freedom
Multiple R-squared:  0.6554,    Adjusted R-squared:  0.6546
F-statistic: 766.4 on 8 and 3223 DF,  p-value: < 2.2e-16
```

Figure 5.11: Sample OLS Output. R Development Core Team (2008). R - A language and environment for statistical computing. R Foundation for Statistical Computing, Vienna, Austria. ISBN 3-900051-07-0, URL http://www.R-project.org.

With GWR, a bandwidth is assigned to each regression point that helps determine how weights will be applied to other properties. The bandwidth can either be fixed (including all properties within some predetermined distance) or adaptive (including some predetermined value of nearest neighbors).

GWR can be used to identify neighborhoods or other submarkets within a jurisdiction. Locational patterns in model diagnostics (e.g., adjusted R^2 values with similar behaviors) and coefficient values) and coefficient values can be used t identify potential market segments. These properties can then be grouped together, which is useful for classification purposes used throughout the assessment process. *Figures 5.12* and *5.13* reveal how patterns can be detected based on model performance.

The bandwidth and weighting specification can have a significant impact on valuation performance, specifically with respect to ratio studies. However, modelers should be careful not to set bandwidths too small, as this could lead to **overfitting**. Overfitting is a phenomenon in modeling which results in the finding of false positive relationships which, in the case of real estate valuation modeling, are not indicative of true market behavior.

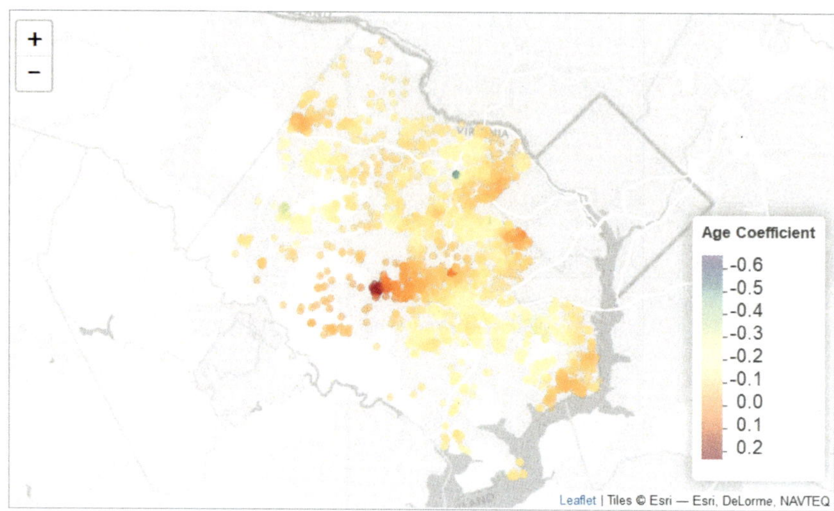

Figure 5.12: Age Coefficient Map. *Source:* Esri and its licensors. Created in R Using the leafletR package. R package citation: Graul, Christian (2016): leafletR: Interactive Web-Maps Based on the Leaflet JavaScript Library. R package version 0.4-0, http://cran.r-project.org/package=leafletR.

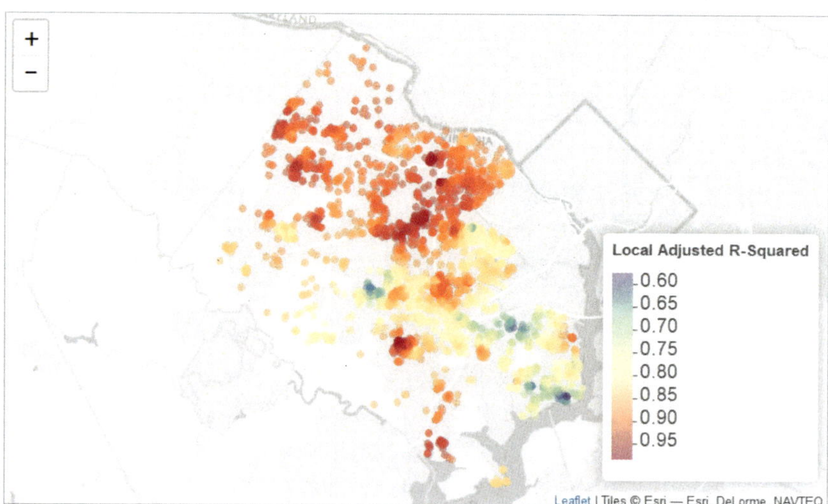

Figure 5.13: Local Adjusted R-Squared Map. *Source:* Esri and its licensors. Created in R Using the leafletR package. R package citation: Graul, Christian (2016): leafletR: Interactive Web-Maps Based on the Leaflet JavaScript Library. R package version 0.4-0, http://cran.r-project.org/package=leafletR.

GWR also reduces the number of models and model variables that are needed for a given area. Say a modeler creates an AVM that looks at (among other variables) how size and neighborhood impact the sale price of a home. The modeler would likely include a variable that represents total living area (in square feet, for this example), and perhaps a dummy variable to denote the specific neighborhood. The coefficient of each neighborhood variable becomes a location adjustment for the base per-square-foot rate. The modeler could also decide to create a separate model for each neighborhood. With GWR, the per-square-foot rate will vary not only at the neighborhood level, but within the neighborhood as well. This promotes efficiency during the modeling lifecycle.

A modeler can also create interaction variables to create spatially varying coefficient estimates. This is accomplished by multiplying a locational dummy variable (e.g., each neighborhood dummy variable) by a quantitative variable (e.g., total living area), resulting in a coefficient estimate for each location.

Incorporating location variables (such as distance or dummy variables) in GWR can be problematic, as it has the potential to introduce bias into model estimates. Modelers should understand the spatial weighting of GWR and adjust bandwidths accordingly, as each has been shown to impact accuracy and uniformity of values.

Figure 5.14 shows varying total living area (TLA) coefficients produced from a GWR model.

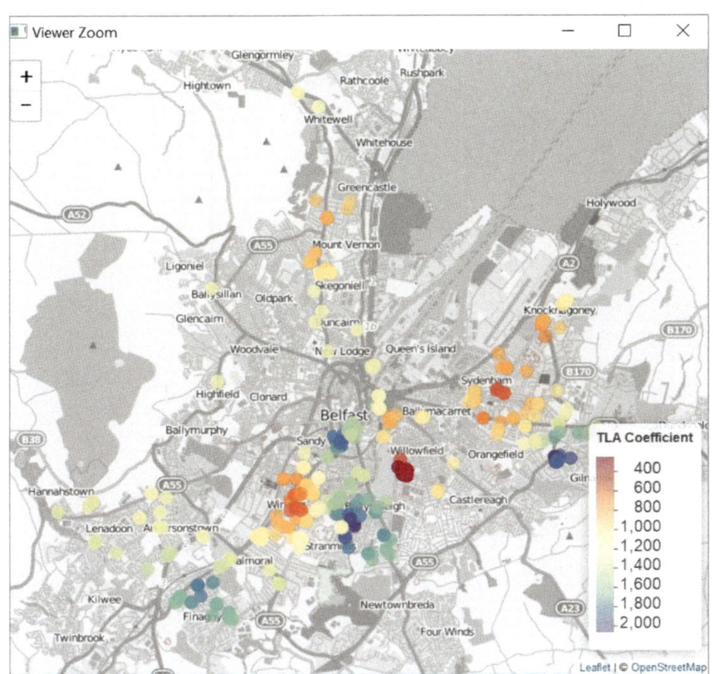

Figure 5.14: Local Adjusted R-Squared Map. *Source:* Esri and its licensors. Created in R Using the leafletR package. R package citation: Graul, Christian (2016): leafletR: Interactive Web-Maps Based on the Leaflet JavaScript Library. R package version 0.4-0, http://cran.r-project.org/package=leafletR.

3.2 | Spatial Interpolation Models

Spatial interpolation utilizes measured locational values to calculate predictions for non-measured locations. During estimation calculations, a higher weight is given to nearer observations. The purpose of such a weighting scheme is to yield estimates that are more congruent with the geographic realities of a region. While a similar methodology is common across a variety of disciplines, in the property tax arena it is most commonly referred to as "response surface analysis" (RSA). RSA can be used to predict property values or other variables, and has demonstrated an ability to estimate sparse pockets of property markets, promoting equity, uniformity, and accuracy of valuations.

RSA is a multi-step process. First, the modeler obtains a location adjustment. One common method to achieve this is by calculating a **Z-score** (the number of standard deviations from the average value) for each observation in a data set, such as on the sale price, price per square foot, or error term. The next step is employing a spatial interpolation process that "smooths" the Z-score points into a "surface."

Figures 5.15 and *5.18* show how spatial interpolation is used to "smooth" points of values on a map into a continuous surface.

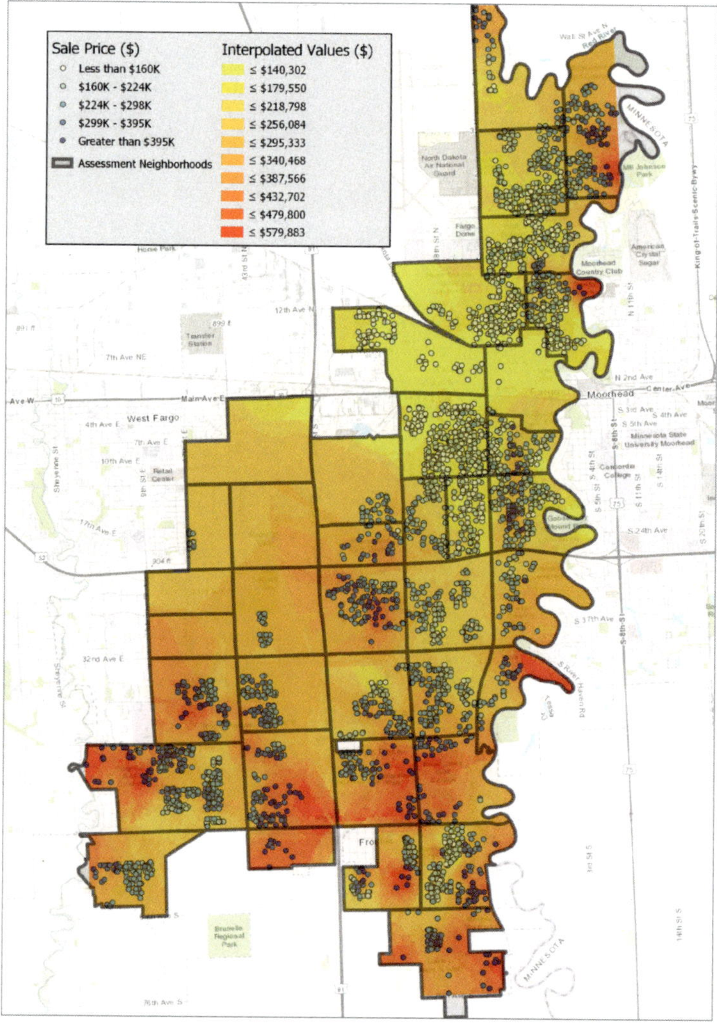

Figure 5.15: Sale Price and Interpolated Sale Price Values. *Source:* Esri and its licensors; City of Fargo, ND, Esri, HERE, Garmin, Intermap, increment P Corp., GEBCO, USGS, FAO, NPS, NRCAN, GeoBase, IGN, Kadaster NL, Ordnance Survey, Esri Japan, Open StreetMap contributors, and the GIS User Community

REAL WORLD APPLICATION: Intuitive RSA

Authors: Paul Bidanset; Billy McCluskey, PhD; and Peadar Davis, PhD, MRICS

Dr. Billy McCluskey of the University of Pretoria is a leading researcher and practitioner in the field of spatial modeling for valuation and property taxation. Much of his recent efforts have revolved around making RSA more intuitive for and accessible to assessors. In an attempt to provide a clearer explanation of the process, following is a step-by-step guide to creating a response surface:

STEP 1: Compile a data set of properties complete with any information you are looking to analyze.

STEP 2: If you do not already have some sort of XY coordinate assigned to each property (e.g., latitude, longitude), geocode each property address using GIS software and append the coordinates to each respective property in your data set.

STEP 3: Choose the variable in a data set of which you wish to make a response surface. This could be a sale price, price per square foot, or any other number of numerical characteristics. Z-scores are commonly calculated to capture deviation. If you chose to perform this calculation on your variable(s), it can be accomplished using common statistical software (e.g., R, SPSS), or even Microsoft Excel.

STEP 4: Now it is time to "smooth" these points into a continuous surface. This can be completed using any number of GIS software, including Esri software (e.g., ArcGIS Pro), opensource software such as R, and more. The offerings for interpolation will vary based on the software, but arguably the most common options will include some sort of kriging function and inverse distance weighting kernels.

STEP 5 (optional): Suppose you wish to use this RSA to assign predictions to properties that were not included in the original analysis (e.g., unsold properties, properties that have transferred after response surface is created)—this can be done simply by first assigning XY coordinates (geocoding) the additional properties, and then conducting a spatial join with the response surface. This process evaluates the location of each additional property and assigns it a value based on where it lies within the response surface. Again, this process is possible using common GIS software.

Recent research by Paul Bidanset, Dr. Billy McCluskey, and Dr. Peadar Davis evaluated how various RSA applications impact uniformity of valuations at the neighborhood level. They found that different interpolation techniques will impact valuation performance. *Figure 5.16* shows just how much different types of RSA valuation approaches can vary with respect to coefficient of dispersion (COD) uniformity statistics.

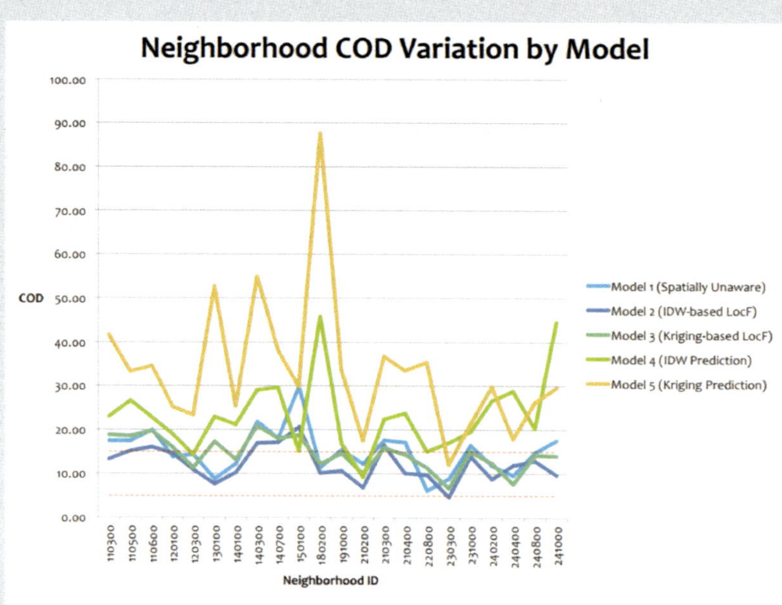

Figure 5.16: Neighborhood COD Variation by Model

The dotted red lines represent the recommended COD acceptability threshold for non-new, single-family homes (between 5.0 and 15.0). Model 2, which employed a location factor adjustment using an inverse distance weighting function, yielded the highest number of "passing" neighborhoods. However, while this produced the best results from an aggregate standpoint, it was still outperformed by other interpolation models (e.g., model 3 in neighborhood 240400). For assessors looking to use RSA for valuation purposes (either location factor adjustments or full-on estimates of value), it is important to explore how each interpolation technique impacts valuation performance at the submarket or neighborhood level in order to optimize overall jurisdiction performance. *Figure 5.17* demonstrates a geovisualized performance of models 2 and 3.

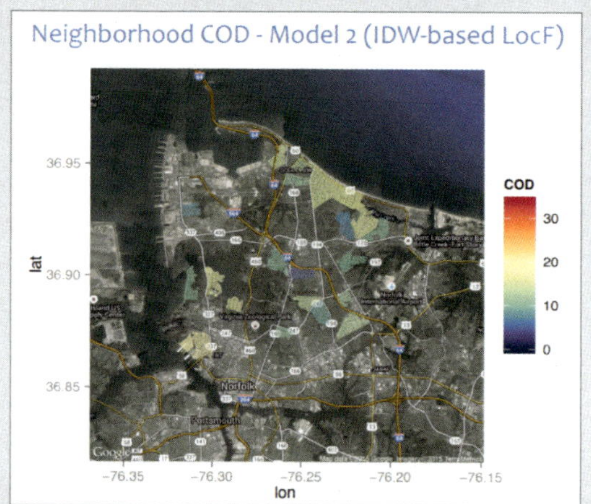

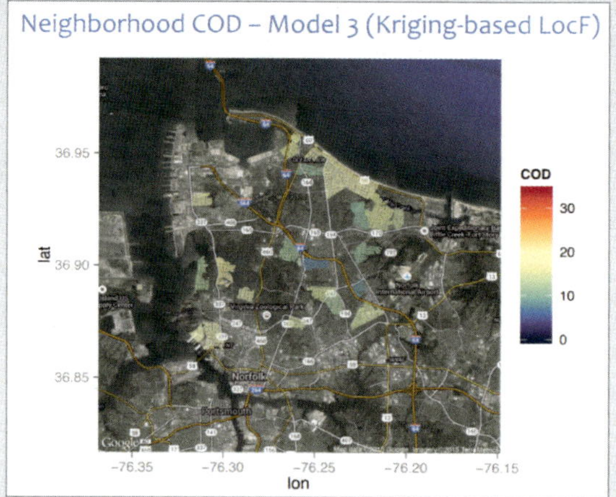

Figure 5.17: Geovisualized Performance of Models 2 and 3. Map data © 2015 Google Imagery © 2015 TerraMetrics

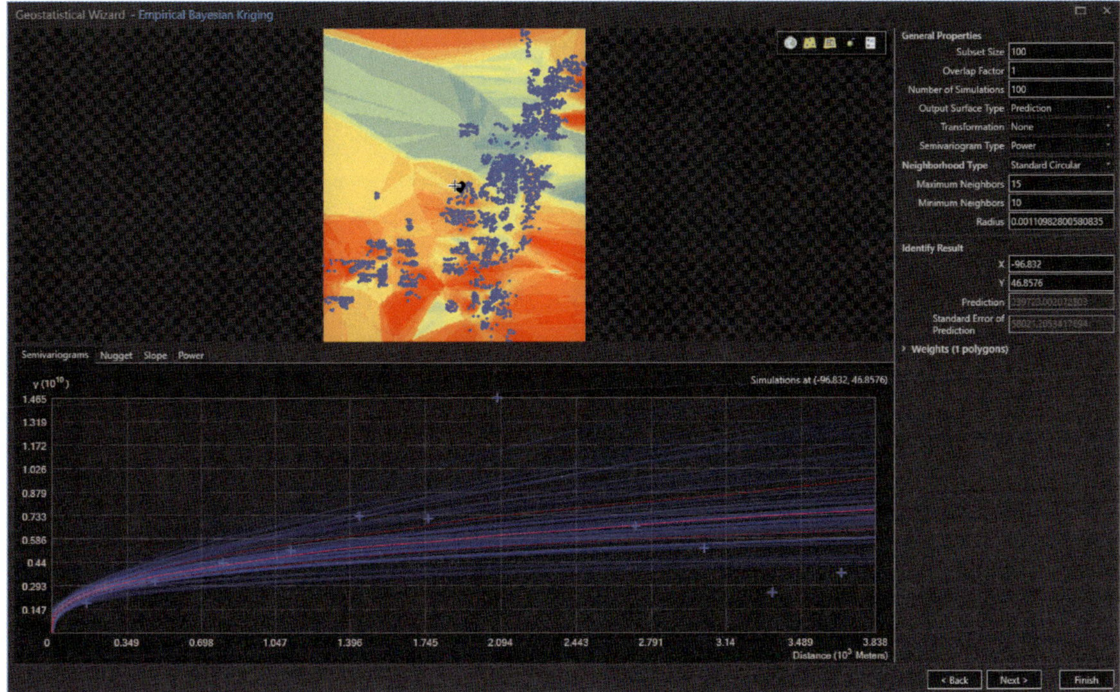

Figure 5.18: Spatial Interpolation in ArcGIS Pro. *Source:* Esri and its licensors; City of Fargo, ND, Esri, HERE, Garmin, Intermap, increment P Corp., GEBCO, USGS, FAO, NPS, NRCAN, GeoBase, IGN, Kadaster NL, Ordnance Survey, Esri Japan, Open StreetMap contributors, and the GIS User Community

Common spatial interpolation modeling techniques include **kriging** and **inverse distance weighting** (IDW). When mapped, this allows for continuous variations in value to be visually identified. In order to obtain a location adjustment for an unobserved location that is covered by this response surface, a spatial join is executed using GIS-capable software. This process can be used to create location adjustment variables able to be used in AVMs, or to independently estimate an overall value.

3.3 | Spatial Regression Models

Spatial regression models have a long history of use within real estate modeling (particularly among econometric research), but they are arguably less prevalent in actual practice among property tax modelers. This is likely just due to preferences of the modeler or software availability limitations, as these modeling approaches have demonstrated a superiority to OLS regression in their ability to estimate prices across locations. Two of the more common spatial regression methodologies include the spatial lag model (SLM) and the spatial error model (SEM). The SLM helps to reduce geographic correlation of residuals

Even if a particular model produces the best results overall, it is important to see how other models perform at the submarket level. With respect to ratio study performance, GWR might work best for one neighborhood, while SLM might work best for another. Each neighborhood should be valued with the model that works best for that neighborhood.

REAL WORLD APPLICATION: New Technologies

Author: Pat O'Connor, President • O'Connor Consulting Inc.

Before the development of response surface software, the concept of wave technology was considered as a replacement for fixed plateau technology of existing neighborhood concepts. Wave technology considers that location value moves slowly over the years (much like wave crests). Unfortunately, assessment jurisdictions usually do not properly maintain their fixed boundary neighborhood boundaries. Neighborhood boundaries do not change with the demographic of the population of property owners. Change is always happening.

From the early tests of a standard house moved around the assessment jurisdictions to creating a mathematical model to represent locational values to current date color response surface maps, the software programs are always improving. The first software for response surfaces used the concepts of color dots that were converted to existing neighborhoods. The programs advanced to prediction models for all the sample geographic area. Modern-day response surface software is capable of developing statewide/province-wide and national agricultural/residential/commercial/industrial locational values. It is a great advancement in location value response surface technology for all appraisers and real estate-oriented professionals, and can be implemented at the local, state, and even national level, as demonstrated in *Figures 5.19, 5.20, and 5.21.*

Over the past 35 years, most of the original developers and participants have retired. This technology has been accepted by a large minority of assessment jurisdictions. Without documentation of these changes, the history will be reduced to a fragmented history in professional papers. The technology will be lost or will have to be re-invented over and over again as professionals look for better statistical models or even just better ways to improve their valuation models.

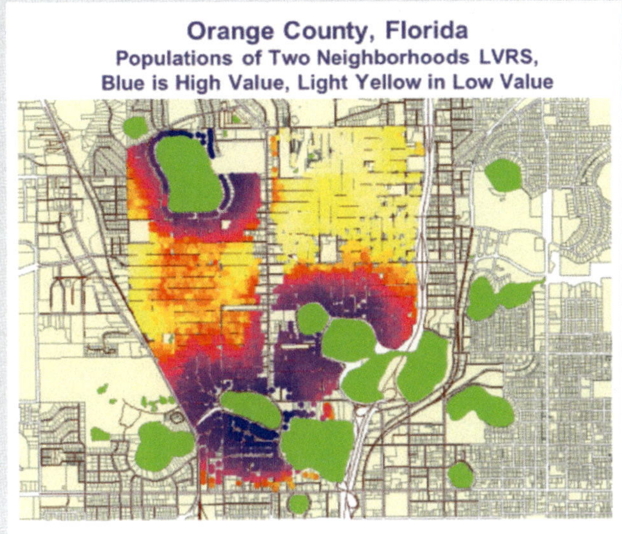

Figure 5.19:
Local Level

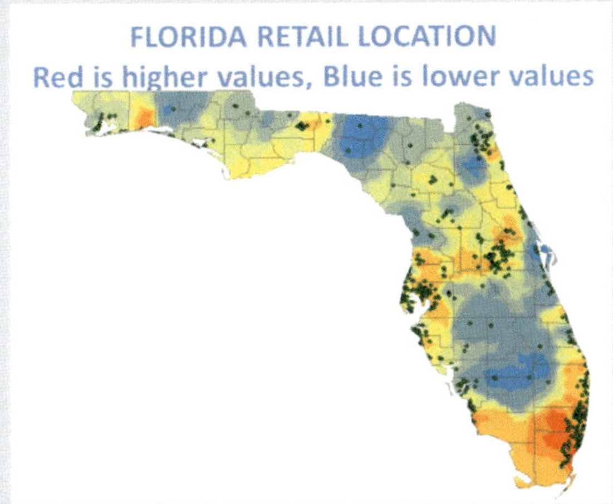

Figure 5.20:
State Level

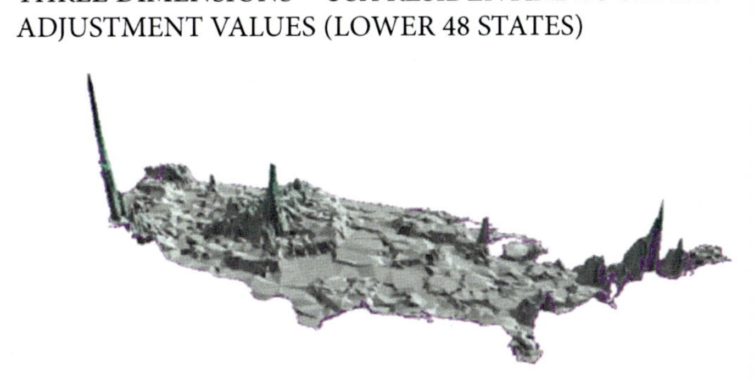

Figure 5.21:
National Level

(i.e., spatial autocorrelation) by adding a spatially lagged dependent variable to the model that incorporates the spatial effects of nearby observations, while the SEM incorporates such effects through the error term.

See Case Study 5.3 ("Value Analysis Using GIS: Building a Spatial CAMA (SCAMA) Model Using Open-Source Software") at this end of this chapter for an illustration of how to diagnose and model spatial effects in CAMA models and how to use free open-source software in spatial modeling.

4 | Conclusion

In appraisal and valuation, GIS makes it possible to properly quantify how a property's value is impacted by its location. Whether through the incorporation of additional variables that describe a property's location, or through the implementation of a more advanced spatial model, such geographic consideration allows assessors to reach more accurate, uniform, and defensible valuations. And as software capabilities and data availability continue to increase, GIS-driven valuation models will only continue to improve, both in terms of efficiency and explanatory power.

Case Study 5.1

Examining the Spatial Relationship Between Environmental Health Factors and House Prices: NO_2 Problem?

Authors: Michael McCord, PhD • Ulster University, Belfast, Northern Ireland
Paul Bidanset; and
Peadar Davis, PhD, MRICS • Ulster University, Belfast, Northern Ireland

Air quality is an environmental externality that has important concerns for health and wellbeing. Unlike the market for most tangible goods, the market for environmental quality does not yield an observable per unit price effect. As no explicit price exists for a unit of environmental quality, this research utilizes the housing market to derive its implicit price and test whether these constituent elements of health and wellbeing are indeed capitalized into property prices and thus implicitly priced in the market place. In order to measure the economic effect of environmental quality, the research integrates geographic information systems (GIS) with housing market data to construct a spatially based analysis of air quality and its association with house prices, using 2,501 sale transactions, providing findings to support local air zone management strategies, noise abatement and management strategies and is of value to the wider urban planning and public health disciplines. Interestingly, if environmental factors such as air, in addition to its recognized impacts on the health and wellbeing, also have an association with the price of residential property across a range of exposure levels (natural consequence) this perhaps illustrates a quadratic trade-off.

METHODOLOGY

The study employs a GWR approach and uses GIS to capture the source pollutants derived from governmental air quality background concentration map database (1km^2 grid) using particulate matter ($PM_{2.5}$ µg/m^3) (*Figure 5.22*) and nitrogen dioxide (NO_2 µg/m3) (*Figure 5.23*) provide concentration levels as total mean concentrations per square kilometer. These air pollution sources were superimposed with *X, Y* coordinated house prices by layering the database of environmental pollutants using GIS, which also incorporated into the GIS mapping to account for proximity to pollutants sources, including the location of major roads and airport runways using distance buffer intervals to provide distance contours to the specific pollutant source. The data was subsequently exported into the statistical package *R* to permit geo-statistical analysis.

RESULTS

The findings reveal an interesting spatial depiction across the coefficient range illustrating both positive and negative effects inferring that low air quality (high levels of NO_2) impact negatively upon prices, with the exception of well-established upmarket housing areas towards the south-south east of the city reflective of the utility trade-off between level of air pollution and desirable

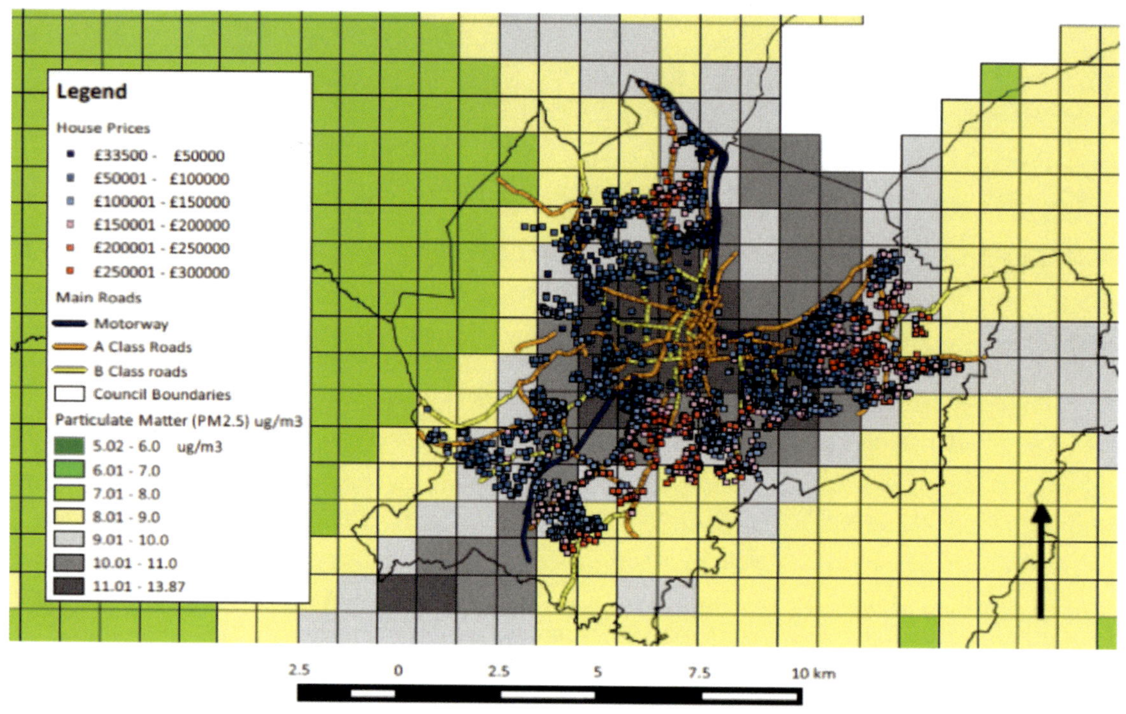

Figure 5.22: House Prices and Particulates Level ($PM_{2.5}$)

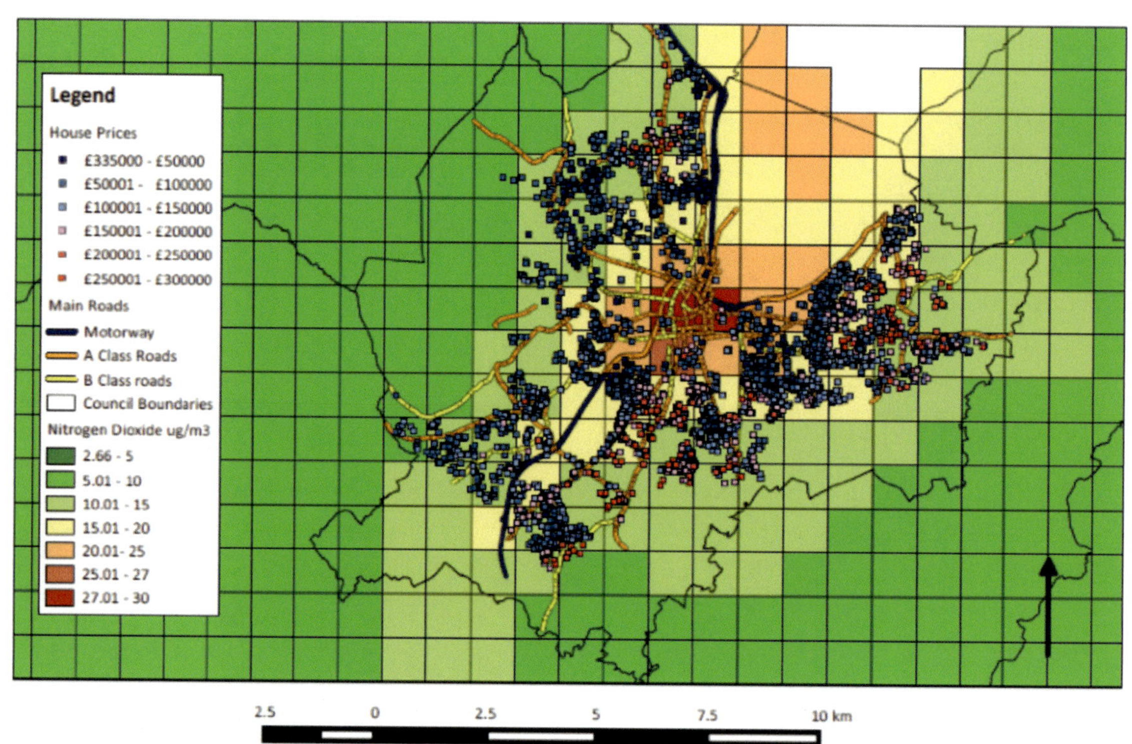

Figure 5.23: House Prices and Nitrogen Dioxide Levels

living locales. What is interesting to note is that the coefficient values are positive at the market median for *high air quality* and *very high air quality* areas which show an increasing positive impact between these ranges—suggesting that higher priced properties 'value' air quality more highly. Combined with the negative signs at low value areas, this suggests a cubic relationship. An interesting finding in this regard pertains to the spatial patterns characterized by higher values in the south west of the city where low air quality negatively impacts on house prices, and high and very high air quality positively impacts on house prices. This pattern is a likely consequence of topography and prevailing winds. Low air quality high value areas remain unaffected whereas the rest of the market appears more anemic, however when examining the spatial variation of the high air quality this appears to impact positively on the market, albeit at different pricing levels, notably advantaging higher value areas (*Figures 5.24-26*).

With regards to the $PM_{2.5}$ air quality parameter, some extreme values exist. For example, the estimates for $PM_{2.5}$ ranging between, 12.01-13, show the coefficient values to differ by -74.7% and +101% in some areas when accounting for high particle matter (poor air quality). What is noticeable however is the direction of the coefficient values when observing the particle matter range. At the 1st quartile, $PM_{2.5}$ shows a positive effect on house prices until the range greater than 11, inferring that higher levels of $PM_{2.5}$ have a negative impact on house prices (12.01-13.0 = -6.27%), whereas lower levels of $PM_{2.5}$ have an increasing positive effect (8.01-9.0 = 14.6%). In addition,

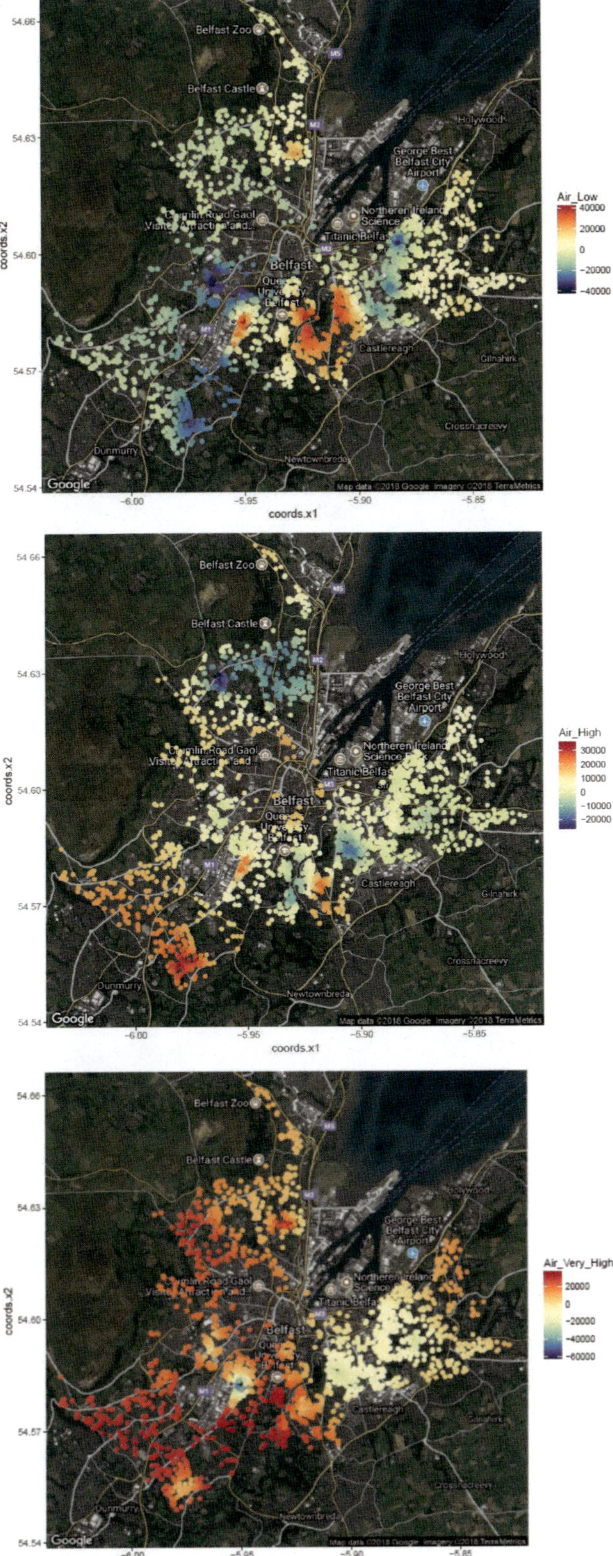

Figure 5.24 (top): NO_2 Air Quality Spatial Representation
Figure 5.25 (middle): NO_2 Air Quality Spatial Representation II
Figure 5.26 (bottom): NO_2 Air Quality Spatial Representation III
Map data © 2018 Google Imagery © 2018 TerraMetrics

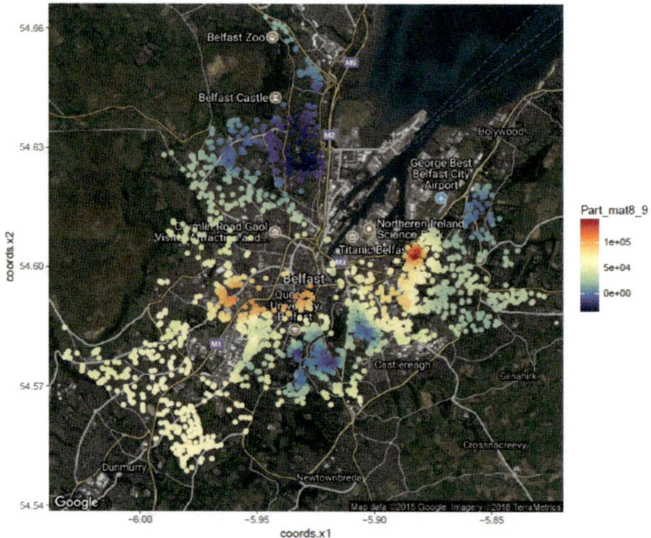

Figure 5.27:
$PM_{2.5}$ Coefficient Spatial Representation I
Map data © 2018 Google
Imagery © 2018 TerraMetrics

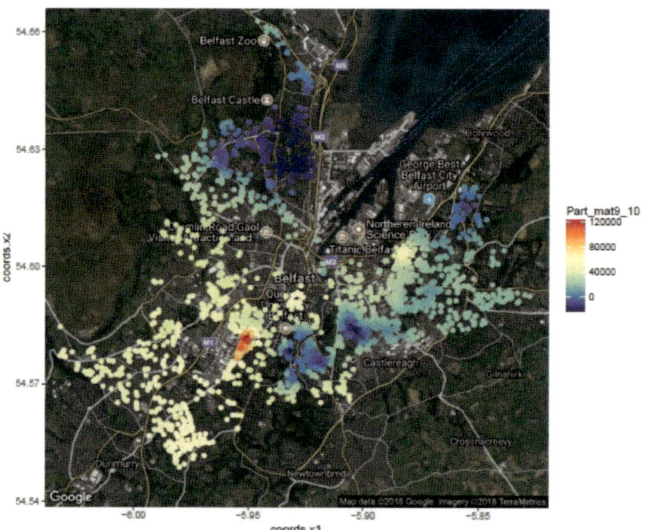

Figure 5.28:
$PM_{2.5}$ Coefficient Spatial Representation II
Map data © 2018 Google
Imagery © 2018 TerraMetrics

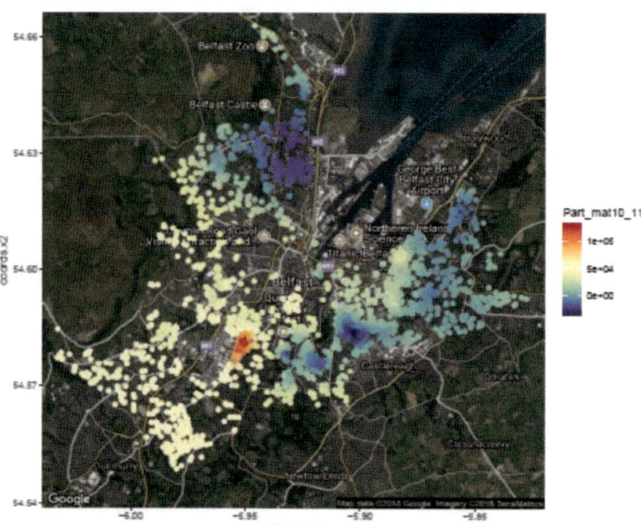

Figure 5.29:
$PM_{2.5}$ Coefficient Spatial Representation III
Map data © 2018 Google
Imagery © 2018 TerraMetrics

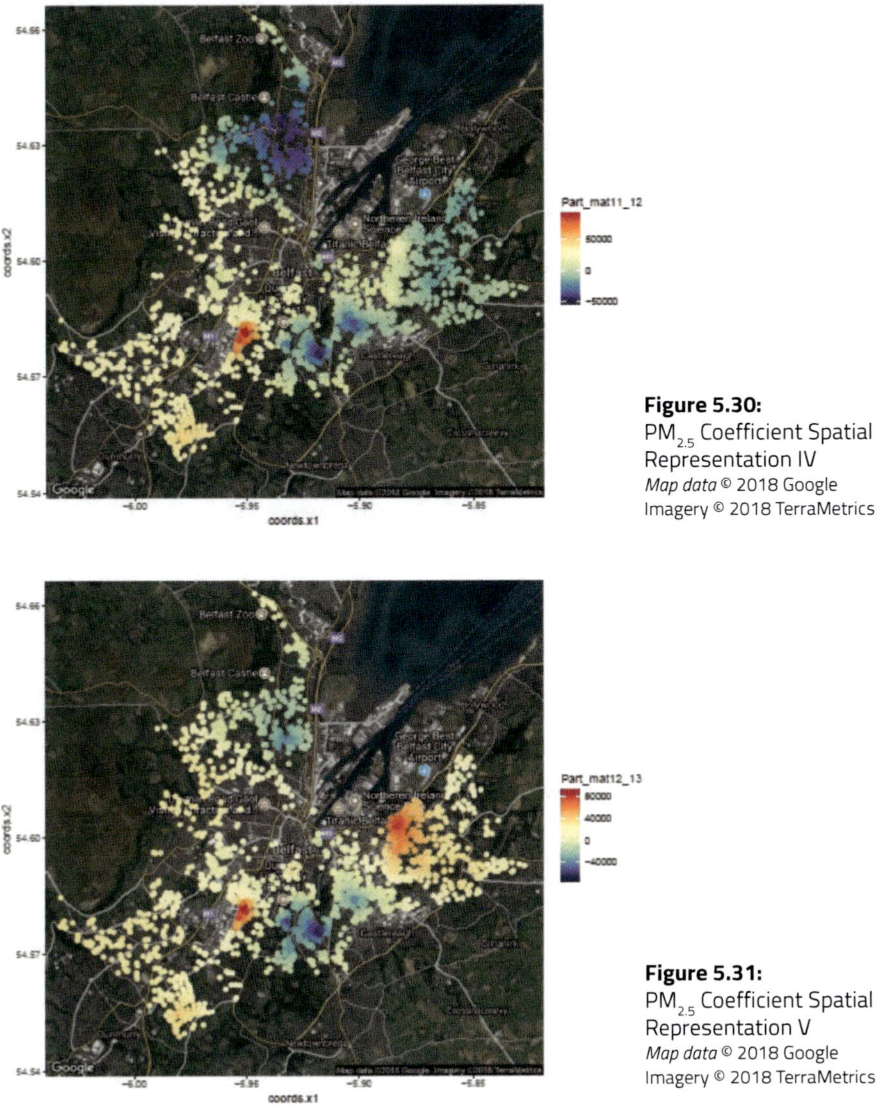

Figure 5.30:
$PM_{2.5}$ Coefficient Spatial Representation IV
Map data © 2018 Google
Imagery © 2018 TerraMetrics

Figure 5.31:
$PM_{2.5}$ Coefficient Spatial Representation V
Map data © 2018 Google
Imagery © 2018 TerraMetrics

the parameter estimates exhibit lower percentage effects on value at the median and 3rd quartile range for the higher levels of particle matter, suggesting that lower levels of $PM_{2.5}$ pollution have a higher positive association with house prices, and higher levels of $PM_{2.5}$, albeit it positive, a lower effect.

CONCLUSIONS

The findings overwhelmingly indicate that air quality seems to have an *asthmatic* effect on property prices. Indeed, nitrogen dioxide consistently illustrates a negative relationship, with $PM_{.5}$ consistently exhibiting an adverse relationship with property pricing. Indeed, the results show a consistent relationship between house prices and air pollution (quality) implying perhaps people living in lower priced property (and arguably of lower socio-economic status) are residing in urban environments which are harmful to health and wellbeing, illustrating a compounding effect between the pollutants, distance, pricing and market structure—the capitalisation effect—to coin a phrase, poor air is bad for your health *wealth* and wellbeing. The 'pockets' of good air in more wealthy neighbourhoods appear to benefit from additional gearing effects, ramping up the house price effects more so than in the general mid-market range—suggesting a cubic profile to the pricing effect. Whilst the results show air quality to have a spatial impact on house prices, they do intimate that across the value distribution this is not a simple effect. Arguably the results capture the variation across the urban landscape and market pricing structure for localised demand and supply tastes—for clean air, and the resulting spatial heterogeneity. For example, apartments located in the city centre region which may suffer from high levels of $PM_{2.5}$ pay more for exclusivity and the trade-off for transport friction costs, city centre living and air quality. They may well enjoy a height advantage and increased wind speed which ameliorates the effects of ground level pollution—a 'z' co-ordinate and distance altitude distance decay statistics would be required to determine this however. Interestingly to the east of the city, the spatial representation of higher levels of particulate matter also show paradoxical results with house prices. Nonetheless, there is a consistent spatial depiction which reveals evidence of market segregation between the north-east of the harbour region and east of the city relative to the south-west area of the Belfast market, where the former across the $PM_{2.5}$ levels are continuously negative, with the latter being positive.

Case Study 5.2

Value Analysis Using GIS: Building a Spatial CAMA (SCAMA) Model Using Open-Source Software

Author: Carmela Quintos, PhD, MAI • Assistant Commissioner, Property Valuation & Mapping Department of Finance, City of New York

INTRODUCTION

This section has two primary objectives: (1) We illustrate steps to diagnose and model spatial affects in CAMA models, and (2) we illustrate spatial modeling using free open-source software. The open-source software we use are developed to solve particular types of statistical problems, are easily distributed, and are developed by academics for easy classroom instruction.

The spatial dimension of the data plays a crucial role when analyzing value where an influence factor is location, such as a neighborhood, town, or district. This is the case as spatial data are characterized by two spatial effects: spatial dependence and spatial heterogeneity. Spatial dependence implies that observations in one location depend on observations in a neighboring location. Spatial heterogeneity, more descriptively referred to as "sub-regional variation", occurs when the processes that drive relationships differ by location. The regression model is misspecified if the influence of location is not correctly captured by the addition of spatial factors.

In practice, it can be difficult to separate the two as spatial dependence can result in spatial heterogeneity. To capture the spatial effect one must use the proper model specification. We discuss two GIS models that are known in mass appraisal: spatial dependence regressions—the spatial loag model (SLM), the spatial error model (SEM), and the spatial autoregressive moving average (SARMA) model—which account for spatial dependence, and geographically eeighted regression models (GWR) which models spatial heterogeneity. There are other models that address spatial effects, for example multi-level (hierarchical) models and random coefficient models to address heterogeneity. This paper focuses on SLM/SEM/SARMA and GWR because they use GIS tools and concepts. We discuss the process of building SLM/SEM/SARMA and GWR models, beginning with exploratory data analysis for spatial effects, diagnostic tests for model calibration, model specification, and tests of fit. We more generally refer to CAMA models that address, through model specification and tests, spatial effects in a GIS environment as Spatial CAMA (SCAMA) models.

The methods are illustrated using land sales data in Brooklyn, New York. The exercise is to predict land prices from physical attributes. We show that adding a spatial dimension using an SLM/SEM/SARMA and GWR model substantially improves model fit.

The software used is free and open-source:

1. GeoDa for SLM/SEM/SARMA modeling. GeoDa is a free, user-friendly software developed for exploratory spatial data analysis, detection of spatial effects, identification of spatial lags, and modeling. Documentation and software can be downloaded from **https://geodacenter.github.io/.**

2. GWR42 to address spatial heterogeneity. GWR4 is a free, user-friendly software for modelling spatially varying regression coefficients. Documentation and software can be downloaded from **https://gwrtools.github.io/gwr4-downloads.html**.

Both software were developed by academics and are widely used in research because they are built specifically to address a particular type of statistical problem and are configured for easy classroom instruction.

SPATIAL DEPENDENCE AND SPATIAL HETEROGENEITY

Regression analysis is the process of using data and summarizing it in a single equation that enables prediction and exploration of hypothesis. The process of finding the equation entails finding patterns in the data. The OLS equation is appropriate if its residuals have a random pattern. A special type of pattern in the residuals would indicate that spatial models are necessary. If residuals show *spatial dependence* or spatial autocorrelation (for example, values in one neighborhood depend on the values in the adjacent neighborhood) then SLM/SEM/SARMA should be used. If, however, residuals show *spatial heterogeneity* (regression coefficients vary by neighborhood) then GWR is an appropriate model specification. Visually, spatial heterogeneity generally refers to the clumps or patches of events in an area, while spatial dependence refers to the processes that create clusters of events. Thus if an area has spatial heterogeneity it may be the result of spatial dependence. Spatial heterogeneity without spatial dependence can be described as, for example, the price regression for neighborhood A having different coefficients from neighborhood B, and the neighborhood prices are independent.

Another differentiating property is that spatial heterogeneity manifests itself in the mean that vary by area, for example regression coefficients which vary by region since regressions are average responses. This is the reason why spatial heterogeneity is also known as a "first-order" spatial effect since statistically, it concerns the first moment (the mean). Spatial dependence on the other hand is a "second-order" effect because it deals with correlations or measures of dependence (statistically, it concerns second moments).

Because spatial heterogeneity may be the result of spatial dependence, the sequence of modeling should check and clear residuals of spatial dependence prior to testing for spatial heterogeneity. A test for spatial heterogeneity is to look at standard statistical tests for heteroscedasticity. Heteroscedasticity is a special case of heterogeneity where the heterogeneity is in the variances (variances are unequal across location). Unless spatial dependence is accounted for, tests for spatial heteroscedasticity will not have the power to determine if the rejection of homoscedasticity is due to spatial dependence or subregional variation.

A recommended modeling workflow:

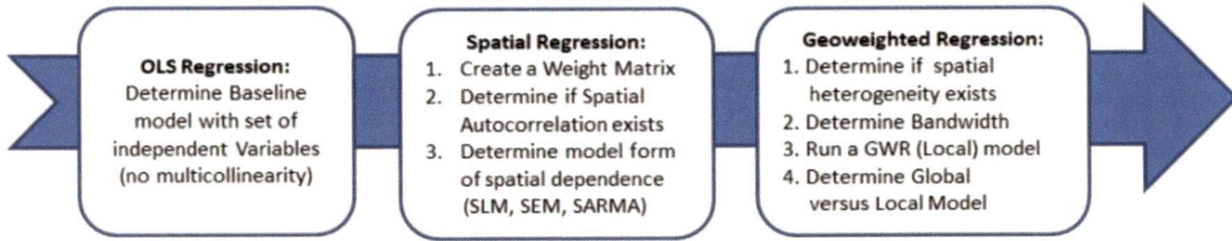

Figure 5.32: Recommended Modeling Workflow

MODELING SPATIAL DEPENDENCE

Consider modeling vacant land prices in Brooklyn, NY. Our dataset consists of 564 arms-length sales from January 2011 to January 2016. Brooklyn is particularly heterogeneous with gentrified pockets in the northwest.

Figure 5.33, left panel, shows sale price per square foot (SP_PSF) and its clustering behavior. The red color indicates higher price per square foot and is clustered in the northwest. This is more clearly seen in the right panel which shows the percentile map of SP_PSF with the highest 1% clustered in the northwest area.

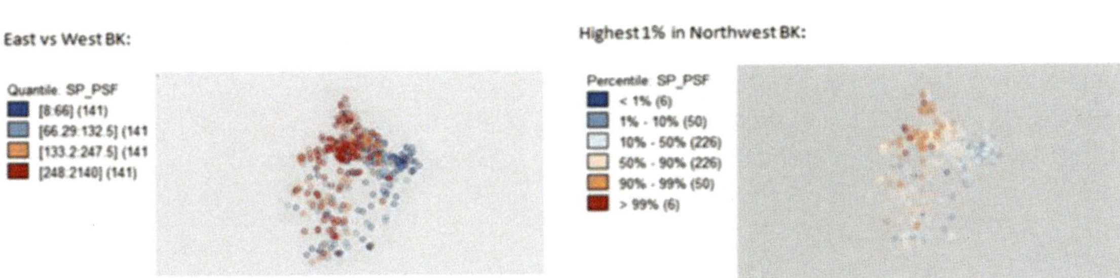

Figure 5.33: Clustering of Sales Price Per Square Foot (in $)

OLS Regression: Baseline Model

Prior to testing for spatial effects, we run the classical OLS regression model. Using stepwise regression, a baseline model of the natural log of price (LNPRICE) is given in *Table 5.3*. The independent variables are natural log of land size (LNSIZE), the usable maximum floor area ratio (USEMAXFAR), the natural log of the distance to the closest public waterfront access (LNWF), the natural log of the distance to the nearest university (LNUNIV), the natural log of the distance to the nearest gallery (LNGAL), the natural log of the restaurant density in parcel area (LNRSTR), a binary variable for a commercially zoned area (ZONECOM), and binary variables, YR11 – Y15, for the year of sale.

The model has four statistically significant locational variables—LNWF, LNUNIV, LNGAL, LNRSTR. There is no indication of multicollinearity from the variance inflation factor (VIF). The VIF assesses how much the variance of an estimated regression coefficient is "inflated" if predictors are correlated. If no factors are correlated, the VIFs will all be 1. The general rule is that VIF's exceeding 4 warrant scrutiny, while VIFs exceeding 10 are signs of serious multicollinearity requiring correction. As shown in *Table 5.3* none of the VIF's exceed 4.

Figure 5.34 summarizes regression residual diagnostics. The residuals are non-normal as indicated by the rejection of the Jarque-Bera test for normality. This is also seen in the OLS residual histogram having a left-skewed tail (high frequency of over prediction). The residuals are also heteroscedastic as indicated by the White Test which is robust to forms of misspecification. Therefore, the OLS regression is misspecified with over prediction and heterogeneity unaddressed in the model specification. The four following section tests whether heterogeneity can be accounted for, at least partially, by correcting for spatial autocorrelation.

Table 5.3: Baseline OLS Model

Baseline OLS Model

Dependent Variable: LNPRICE

Number of Observations Read	564
Number of Observations Used	564

Analysis of Variance

Source	DF	Sum of Squares	Mean Square	F Value	Pr > F
Model	12	562.21121	46.85093	72.21	<.0001
Error	551	357.48146	0.64879		
Corrected Total	563	919.69267			

Root MSE	0.80547	R-Square	0.6113
Dependent Mean	12.73913	Adj R-Sq	0.6028
Coeff Var	6.32282		

Parameter Estimates

Variable	DF	Parameter Estimate	Standard Error	t Value	Pr > |t|	Variance Inflation
Intercept	1	7.35593	0.91645	8.03	<.0001	0
LNSIZE	1	1.11030	0.04930	22.52	<.0001	1.15087
USEMAXFAR	1	0.28410	0.04306	6.60	<.0001	1.40195
LNWF	1	-0.12203	0.04291	-2.84	0.0046	1.33565
LNUNIV	1	-0.23219	0.05623	-4.13	<.0001	1.91795
LNGAL	1	-0.18520	0.05389	-3.44	0.0006	1.70746
LNRSTR	1	0.19606	0.05519	3.55	0.0004	1.86744
ZONECOM	1	-0.43767	0.20363	-2.15	0.0320	1.11373
YR11	1	-0.55794	0.14504	-3.85	0.0001	1.98813
YR12	1	-0.46996	0.13244	-3.55	0.0004	2.41050
YR13	1	-0.38461	0.13514	-2.85	0.0046	2.27941
YR14	1	-0.10229	0.13074	-0.78	0.4343	2.44286
YR15	1	0.02261	0.13163	0.17	0.8637	2.38101

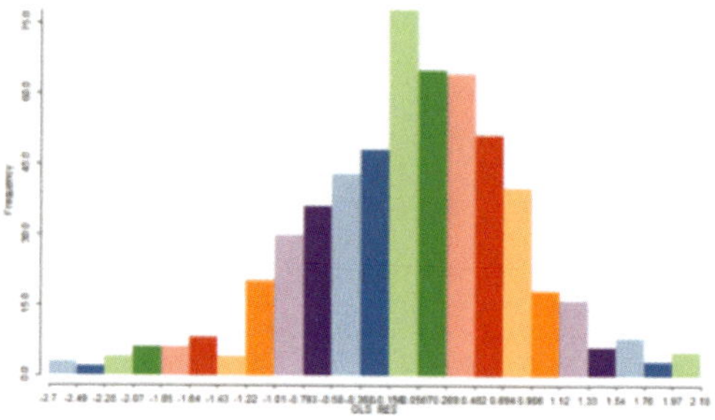

Figure 5.34: OLS Residual Diagnostics

OLS Residual Diagnostics

REGRESSION DIAGNOSTICS

TEST ON NORMALITY OF ERRORS

TEST	DF	VALUE	PROB	Result
Jarque-Bera	2	24.2128	0.00001	Non-Normal Residuals

DIAGNOSTICS FOR HETEROSKEDASTICITY

RANDOM COEFFICIENTS

TEST	DF	VALUE	PROB	Result
Breusch-Pagan test	12	23.1158	0.02676	Heteroscedastic Residuals
Koenker-Bassett test*	12	16.9503	0.15148	Homoscedastic Residuals
White**	90	564	0.0000000	Heteroscedastic Residuals

* Robust Test to departures from the Gaussian or Normal hypotheses

** Robust to Forms of Misspecification

Spatial Regression: SLM, SEM, and SARMA

Spatial autocorrelation is the property that measures the degree to which near and distant neighborhoods (or locations) are related. How the neighborhoods relate to each other is defined by a weight matrix, ***W***.

1. **Creating the Weight Matrix**

   The two most common ways that neighborhood relation is defined is by contiguity (common boundary) and distance.

   - In GeoDa, the "Weights File Creation" manager clearly specifies these two options (*Figure 5.35*).

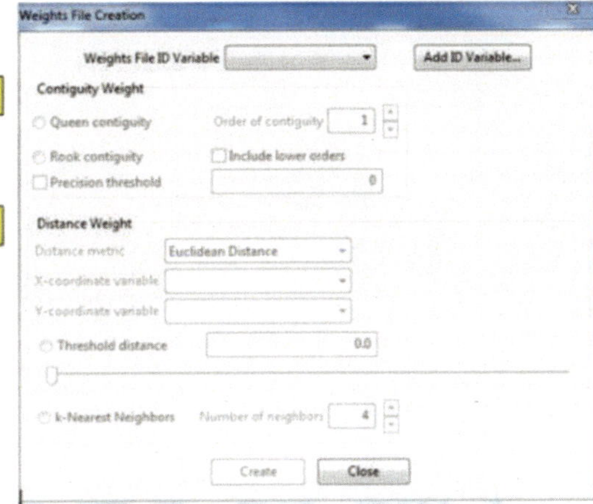

Figure 5.35: Weights File Creation

   - As in a chess board, **Rook Contiguity** means that the neighborhoods share common boundaries that are similar to a rook movement. **Queen contiguity** is like the movement of a Queen with common boundaries and vertices (*Figure 5.36*).

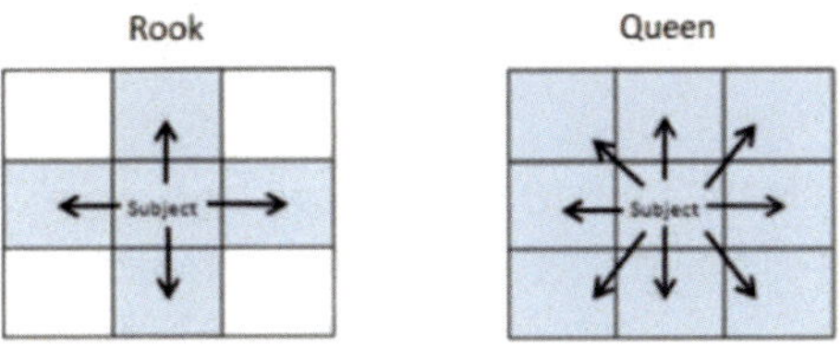

Figure 5.36: Rook vs. Queen Contiguity

   - **Distance-based weight matrices** can specify neighbors based on the k-nearest neighbors or based on a specified distance (*Figure 5.37*).

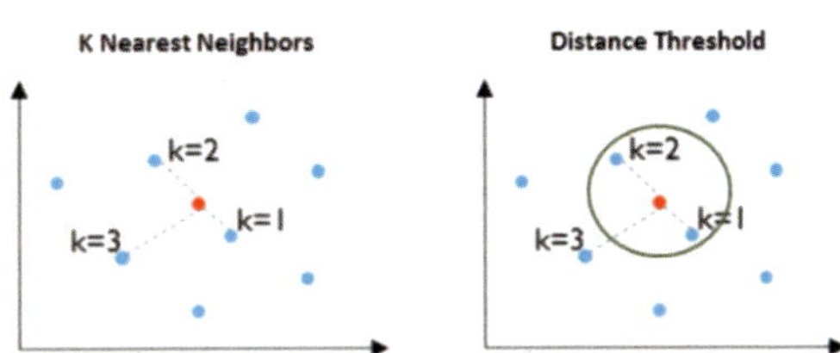

Figure 5.37: Distance-Based Weight Matrices

   The k-nearest neighbors will select the closest k-neighbors regardless of distance. The distance-based method draws a radius in feet as specified by the user and will select every point in the radius as a neighbor.

   If data is sparse the k-nearest neighbors is preferable since a small radius may not pick up any neighbors. To prevent zero neighbors, GeoDa has a default threshold distance which ensures that every observation has at least one neighbor.

   For the vacant land sales data, we start with the default threshold distance of 4,983.77 feet (*Figure 5.38*).

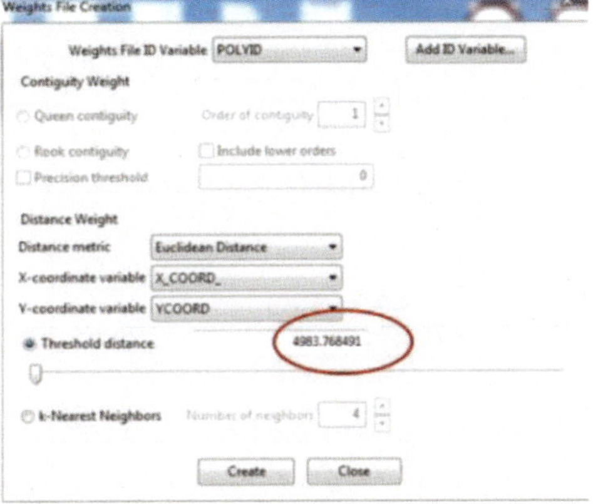

Figure 5.38: Threshold Distance

Connectivity Histograms

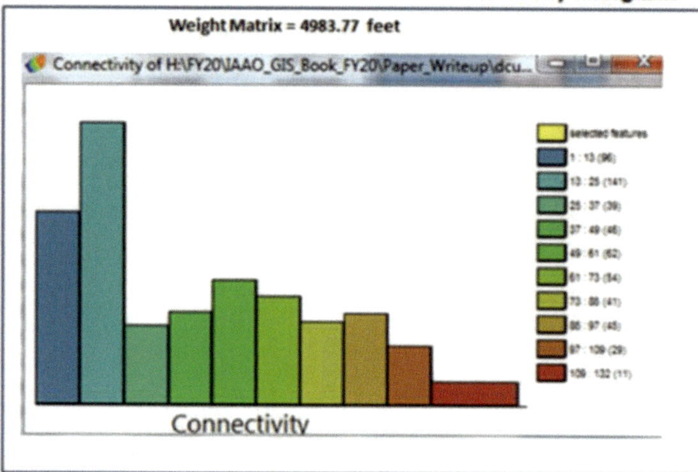

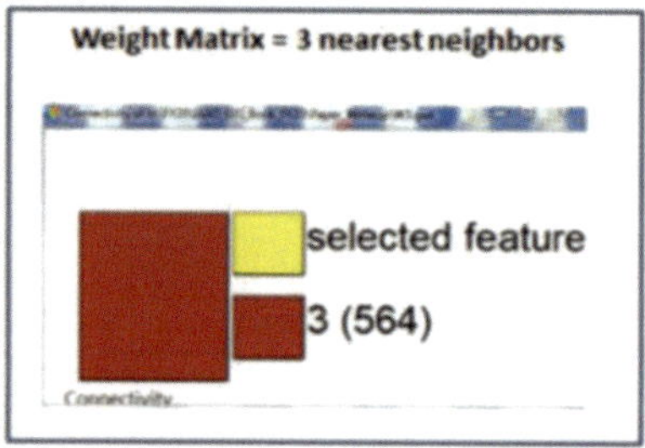

Figure 5.39: Connectivity Histograms

- GeoDa also provides the **connectivity histogram** which shows the distribution of the number of points that connected to a parcel. Too many points captured in a radius will cross several areas and may not capture heterogeneity well.

  In the connectivity graphic (*Figure 5.39*), the histogram on the left panel shows the number of parcels in parentheses by the number of other parcels they connect with. For example, there are 96 parcels that connected with 1-13 neighboring points. The maximum number of points connected to a parcel is 132 which will clearly locate across very different areas.

  Alternatively, we specify the neighbors to be k = 3 regardless of distance. The choice of 3 follows the rule of selecting comparable properties—three comps is the minimum required to get a median sale price. The k-nearest neighbor selection in GeoDa is illustrated in *Figure 5.40*.

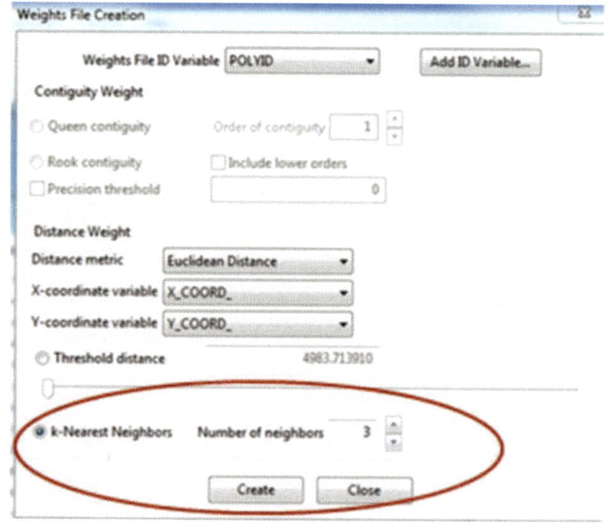

Figure 5.40: K-Nearest Neighbors

  To confirm, we look at the connectivity histogram in *Figure 5.39* right panel. It shows that all our 564 observations are associated with exactly 3 of its nearest neighbors.

2. **Tests for Spatial Autocorrelation –** We perform cluster analysis using global Moran's I, LISA significance maps and LISA cluster maps in GeoDa.

   - **Global Autocorrelation -** Moran's I is a test statistic for spatial autocorrelation. It tests if prices cluster or not given their locations to other points. The mean and variance of prices are calculated. The deviation from the mean is calculated for each point, then the deviation is multiplied with connected neighboring points, as specified by a weight matrix, to create a cross-product. If neighboring price deviations have high or low cross-products, then there is clustering. The test result is interpreted like a correlation result. This gives a global result for the entire research area.

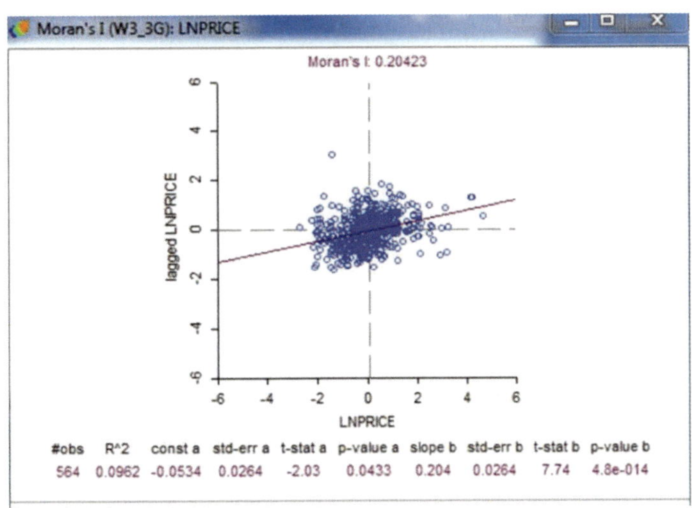

Figure 5.41: Moran's I

Using the weights file in the previous section with k=3 nearest neighbors, Moran's I is the slope coefficient on the standardized weighted cross-products. The slope or correlation coefficient is .20423 and is significant with a t-statistic of 7.74 and a p-value less than 1%. The slope coefficient measures the correlation between LNPRICE and its connected neighbors as determined by the weight matrix with k=3 nearest neighbors. Its significance is indicative of spatial autocorrelation.

We calculate Moran's I for alternative values of k in *Figure 5.42*. The highest Moran's I values occur at k = 4 and k = 6. However, k=6 has a higher t-statistic so we use this specification since it produces the strongest evidence of spatial autocorrelation.

Global Moran's I for Different Values of K

k	Moran's I	t-stat	p-value
3	0.204	7.74	<.0001
4	0.226	9.17	<.0001
5	0.225	9.69	<.0001
6	0.226	10.1	<.0001
7	0.216	9.66	<.0001
8	0.211	9.74	<.0001
9	0.205	9.71	<.0001
10	0.207	10.1	<.0001
11	0.199	9.79	<.0001
12	0.200	9.85	<.0001
13	0.204	10.1	<.0001
14	0.204	10.3	<.0001
15	0.205	10.5	<.0001

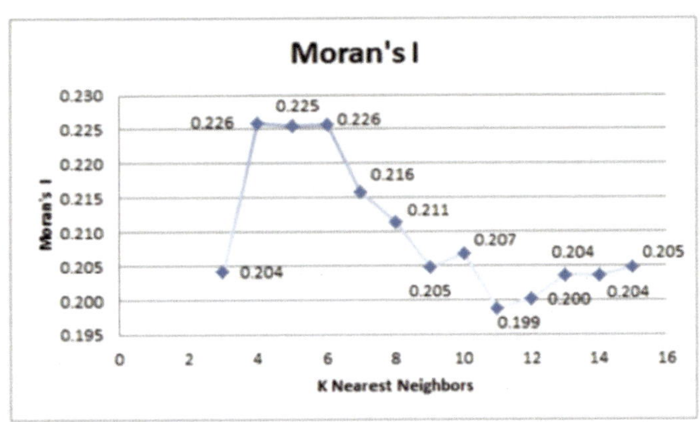

Figure 5.42: Global Moran's I for Different Values of **K**

From the significance of Moran's I, we know there is clustering in the entire area but we do not know where or how the clustering looks. To investigate further, we look at local indicators of spatial association (LISA). These test not only global clustering but it also shows the presence of significant spatial clusters by region.

- **Local Autocorrelation –** The LISA significance map shows significant results by region. Even if the Global Moran's I shows significance for the entire area, only some regions show significant clustering. *Figure 5.43* is the LISA significance map for LNPRICE with k=6 nearest neighbors. There is significant clustering in small portions of the northwest and northeast.

  The LISA cluster map shows how prices cluster. The red color shows "High-High" clustering, meaning high-priced neighborhoods are connected with high-priced neighborhoods. The blue color shows "Low-Low" clustering, meaning low-priced neighbors connect with other low priced neighborhoods. *Figure 5.44* is the LISA cluster map for LNPRICE with k=6 nearest neighbors. The significant clustering in the northwest portion is for high priced neighborhoods, while the significant clustering in the northeast is for low priced neighborhoods.

  From the LISA analysis, we conclude that there is statistically significant clustering in Brooklyn land prices. Higher prices are clustered in the northwest area, while lower prices are clustered in the northeast.

LISA Significance Map for Local Clustering

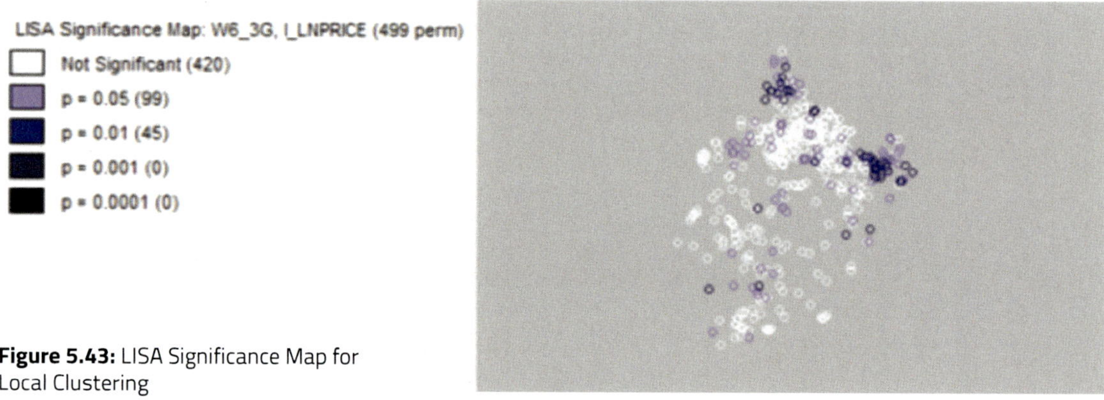

Figure 5.43: LISA Significance Map for Local Clustering

LISA Cluster Map

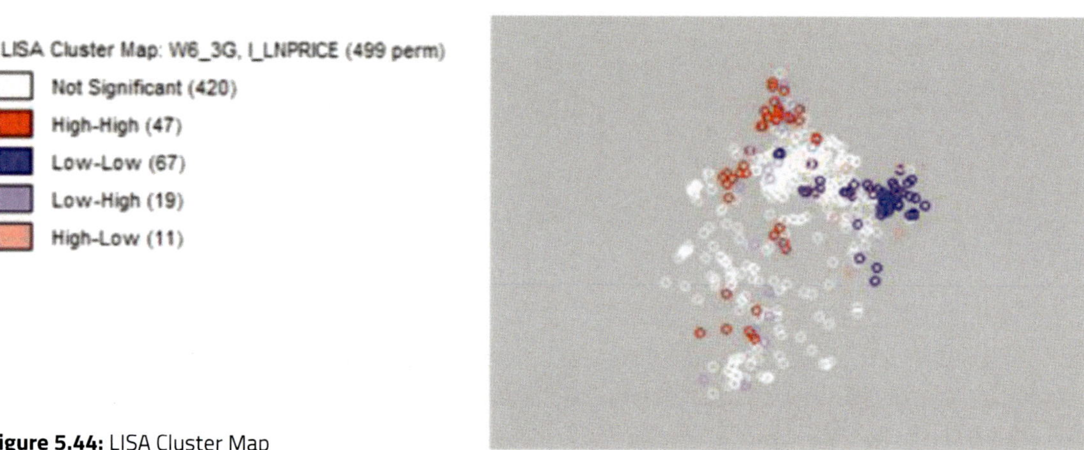

Figure 5.44: LISA Cluster Map

3. **Incorporating Clustering Behavior in Land Price Models: Spatial Regressions –** We know from Moran's I that there is significant spatial autocorrelation. We consider three spatial regression models in this section. Spatial lag models (SLM) includes the neighbors weighted prices as an independent variable and captures measured spatial dependence in the regression coefficient. Spatial error models (SEM), on the other hand, captures spatial dependence in the errors so the influence of spatial dependence is unmeasured. Spatial autoregressive moving average (SARMA) models combine SLM and SEM.

SLM, SEM and SARMA

SLM is a regression equation with a spatial lag as an explanatory variable. The spatial lag is constructed by defining which surrounding points, and how the surrounding points, should influence price in a particular location. For example, are prices in one neighborhood correlated with the movement in price of its surrounding neighbors? If there is a correlation then the surrounding neighborhood's price should be included as an explanatory variable, which is what we refer to as the spatial lag.

The spatial lag is the NxN weight matrix $\boldsymbol{W}$ multiplied by the $Nx1$ vector of prices y where $N = 564$ is the number of observations. The matrix product $\boldsymbol{W}y$ is the weighted average price of the connected neighborhoods. In our case $\boldsymbol{W}y$ is the weighted average of the 6 connected neighborhoods with $\boldsymbol{W}$ constructed from the k=6 nearest neighborhoods. The weighting decreases with the distance of the kth neighbor.

The SLM equation is the OLS equation plus a spatial lag as an independent variable:

$$y = \rho(\boldsymbol{W}y) + x\beta + e$$

where y is a $Nx1$ vector of observations, $\boldsymbol{W}y$ is an $Nx1$ vector of spatial lags, ρ (Rho) is the spatial autoregressive coefficient, x is an NxK matrix of explanatory variables, β is a $Kx1$ vector of regression coefficients and e is the error.

OLS is a special case when $\rho = 0$. Thus a test of an SLM model is the significance of ρ.

SEM includes a spatial autoregressive effect in the error term:

$y = x\beta + e$

$e = \lambda \boldsymbol{W}e + u$

where $\boldsymbol{W}$ is the weight matrix that averages the neighboring errors. Here OLS is a special case when $\lambda = 0$. Thus a test of a SEM model is the significance of λ.

SARMA combines the lag effects of SLM and SEM:

$$y = \rho(\boldsymbol{W}y) + x\beta + e$$

$$e = \lambda \boldsymbol{W}e + u$$

A test of SARMA is a joint test on the significance of ρ and λ.

SLM captures the influence of measurable spatial (locational) influence. In contrast, SEM and SARMA capture the spatial effect but it is unmeasured (i.e. it is in the error) due, most likely, to omitted variables or functional form misspecification. In the latter case, modeling must address the misspecification rather than using SEM or SARMA as the final model even though spatial dependency is captured.

SCAMA Modeling of Spatial Autocorrelation

The regression tab in GeoDa has the option to run an OLS, SLM or SEM model. Before we run SLM or SEM, we re-run the OLS model with a test for spatial dependence in the residuals using the weight matrix with k=6 nearest neighbors. The regression dialog box has the OLS, SLM and SEM options (*Figure 5.45*):

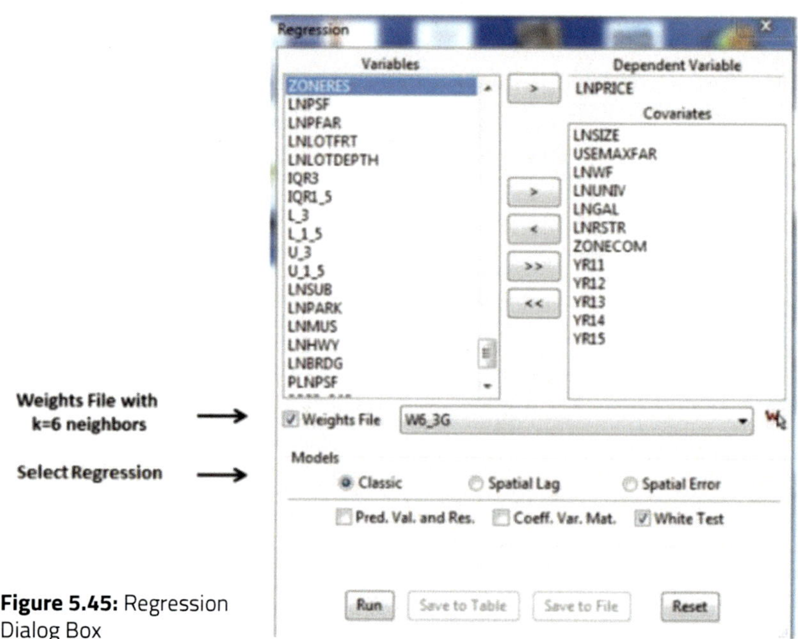

Figure 5.45: Regression Dialog Box

SLM MODEL
Test for Spatial Dependence
Weight Matrix with k=6 Nearest Neighbors

DIAGNOSTICS FOR SPATIAL DEPENDENCE
FOR WEIGHT MATRIX : W6_3G
(row-standardized weights)

TEST	MI/DF	VALUE	PROB
Moran's I (error)	0.1133	5.4658	0.00000
Lagrange Multiplier (lag)	1	25.1824	0.00000
Robust LM (lag)	1	5.5085	0.01892
Lagrange Multiplier (error)	1	22.8274	0.00000
Robust LM (error)	1	3.1535	0.07577
Lagrange Multiplier (SARMA)	2	28.3359	0.00000

Table 5.4: OLS Baseline Model Test for Spatial Dependence

The OLS baseline model coefficients and diagnostics for non-normality and heteroscedasticity were discussed previously in *Table 5.3*. We focus on the last section which tests for spatial autocorrelation and is repeated here as *Table 5.4*.

Moran's I on the error has a spatial correlation coefficient of .1133 which is significant at the 1% level. Moran's test tells us there is spatial dependence in the error but it does not tell us whether we should use SLM, SEM or SARMA. The Lagrange Multiplier (LM) tests and its Robust counterparts do.

Anselin (2005) gives the decision rules to determine the form of spatial dependence. First, use the LM tests on lag and error to separate out the source of spatial dependence. If one is significant, lag or error, proceed to run that model. If all are significant, check the robust LM tests. If one of the robust LM test is significant, lag or error, proceed to run that model. If both the robust LM tests are significant then choose the test with the largest value as an indicator for the spatial model.

Below summarizes the spatial regression decision process from Anselin (2005), page 199:

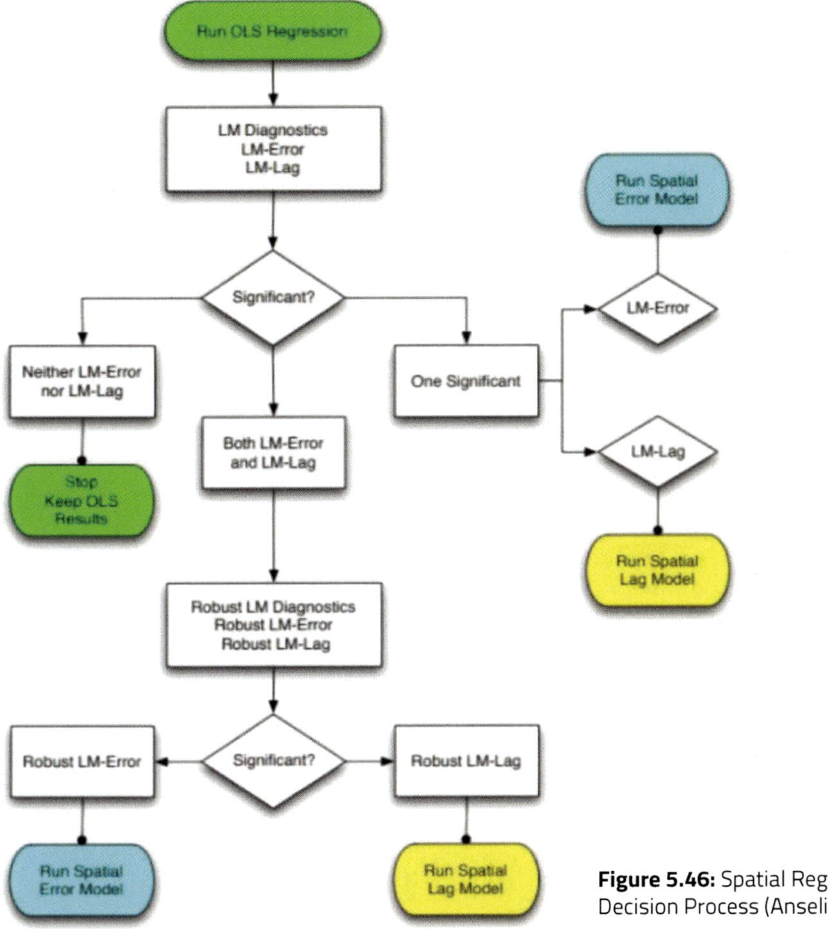

Figure 5.46: Spatial Regression Decision Process (Anselin, 2005)

In our case, the LM-Error, LM-Lag and LM-SARMA tests are significant at the 1% level. Proceeding down the chart, we look at the robust LM-error and robust LM-lag tests, both are significant. However, the robust LM-lag has a test value of 25.1824 which is higher than the LM-error test value of 22.8274. We select the SLM model because of the higher robust LM test value. Note that if we selected SEM we have unmeasured spatial dependence which would mean misspecification due to omitted variables or functional form.

SLM Model for Land Prices in Brooklyn

We modify the OLS baseline model by adding a spatial lag term as an explanatory variable. Compare the OLS baseline diagnostic results of *Table 5.4* to the SLM tests below:
Moran's I on the error is -.0304 which is not significant at the 1% level. Similarly, all the LM tests and its robust counterparts find no significant spatial dependence.

SLM MODEL
Test for Spatial Dependence
Weight Matrix with k=6 Nearest Neighbors

DIAGNOSTICS FOR SPATIAL DEPENDENCE
FOR WEIGHT MATRIX : W6_3G
(row-standardized weights)

TEST	MI/DF	VALUE	PROB
Moran's I (error)	-0.0304	-0.9613	0.33641
Lagrange Multiplier (lag)	1	1.1485	0.28386
Robust LM (lag)	1	0.0062	0.93736
Lagrange Multiplier (error)	1	1.6447	0.19969
Robust LM (error)	1	0.5023	0.47848
Lagrange Multiplier (SARMA)	2	1.6508	0.43805

Table 5.5: SLM Model Test for Spatial Dependence

The LISA significance map and LISA cluster map on the residuals find no pattern of spatial dependence. There is no clustering in residuals in the northwest or northeast as previously seen in *Figure 5.43* and *5.44* for prices. Therefore, with SLM there is no significant spatial dependence that remains unmeasured.

Finally, a comparison of the model fit for OLS and SLM is summarized in *Table 5.6*.

SLM has a higher ddjusted R-squared and lower AIC and Schwartz criterion indicating overall a better predictive model. SLM does not have significant spatial dependence so the heteroscedasticity that is still unmeasured is due to omitted variables or functional form misspecification. We can eliminate the source of heteroscedasticity as coming from spatial dependence.

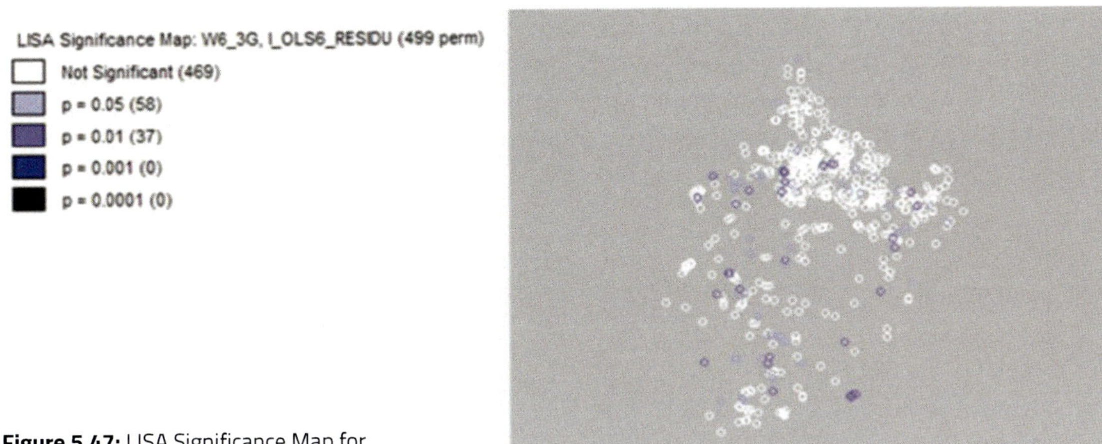

Figure 5.47: LISA Significance Map for SLM Residuals

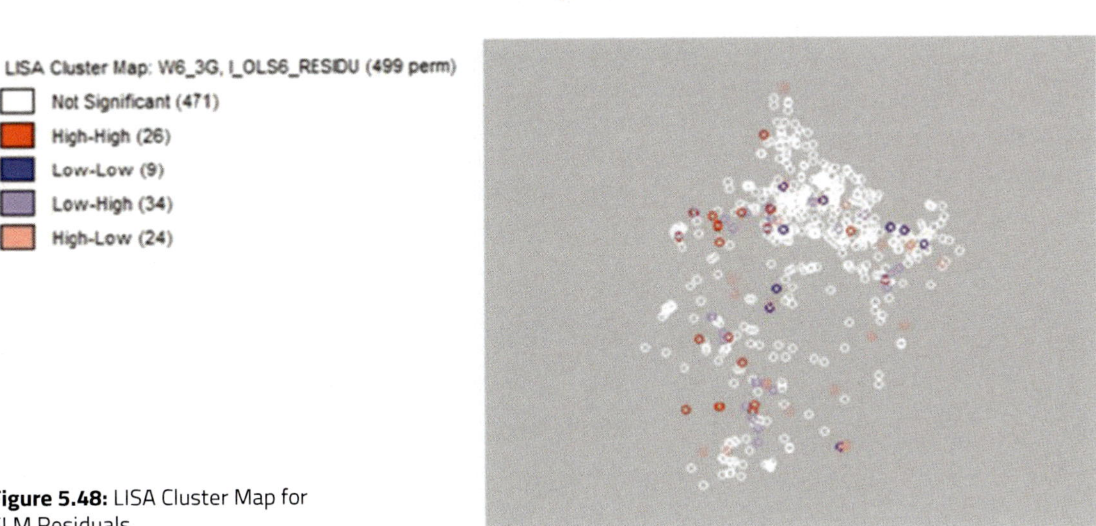

Figure 5.48: LISA Cluster Map for SLM Residuals

Table 5.6: Summary SLM versus OLS Model

(Red indicates better statistical performance)

DIAGNOSTIC TESTS	OLS	SLM
Model Fit		
Adjusted R-Square	60%	63%
AIC	1369.4	1330.36
Schwarz Criterion	1426..75	1391.05
Spacial Dependence	Yes	No
Heteroscedasticity	Yes	Yes

MODELING SPATIAL HETEROGENEITY

SLM and OLS are *global* models in that there is one equation fit across geography. In contrast, GWR is a *local* model because it fits a regression equation to every geographic feature – for example a regression at every point, or a regression at every neighborhood centroid. This is called spatial heterogeneity or spatial non-stationarity since, depending on location, different regression coefficients are used to predict price. Non-stationarity arises from intrinsic local differences so that global statements of spatial behavior are not possible.

The traditional OLS regression model with k independent variables is,

$$\mathbf{y} = \beta_0 + \sum_{j=1}^{K} \beta_j x_j + u,$$

which provides one set of regression coefficients, $\beta_0, \ldots, \beta_k$. In contrast, a local GWR regression provides different regression estimates for different geographical locations,

$$y_i = \beta_{oi} + \sum_{j=1}^{k} \beta_{ji} x_{ji} + u_i,$$

where $\beta_{0i}, \ldots, \beta_{ji}$ are estimated for each independent variable j and each geographical location i. GWR4 allows estimation of a mixed GWR where the modeler can specify which variables are global versus local (allowed to vary by region).

The local parameters $\beta_{0i}, \ldots, \beta_{ji}$ are estimated by a weighted least squares procedure. The weights at each point location are defined by a distance decay function. The weighting functions are called "kernels". The most common kernels are the Gaussian and the bi-square function. GWR4 kernel tab shows the choices (*Figure 5.49*):

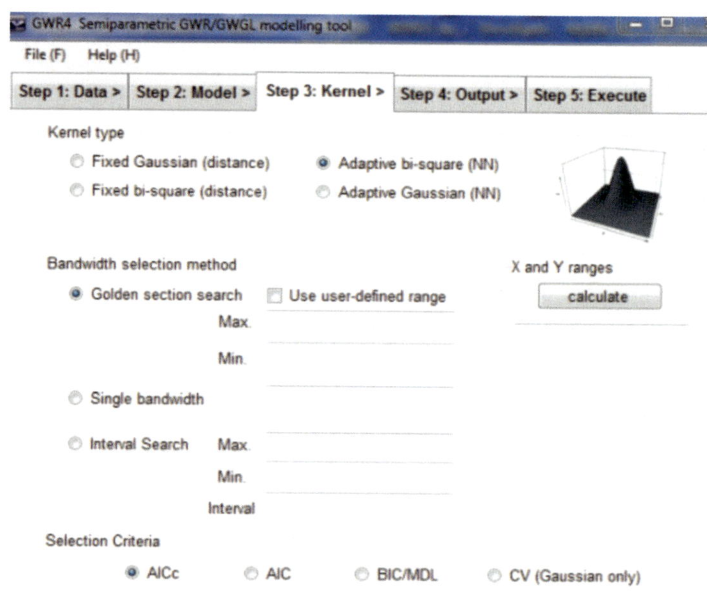

Figure 5.49: GWR4 Kernel Tab

The kernels or weighting functions are defined by two parameters: the distance to the subject point and a so-called "bandwidth" which controls the degree of distance decay. If the bandwidth is large then a relative higher weight is assigned to more remote regions; if the bandwidth is small, then the smoothing effect is low so relative lower weights will be assigned to more remote regions.

The fixed kernels are defined by a fixed bandwidth which works if the data is evenly spaced. The problem with fixed bandwidth is that undersmoothing occurs in areas with small observations, while oversmoothing occurs in areas with a high density of points. An illustration of a fixed kernel, or weighting function wij, is given below. The width of the bell curve from the data point x is the bandwidth.

The adaptive kernels are defined by spatially varying bandwidths. The kernels (bandwidths) are smaller in regions with high density of points and are larger in regions with low density of points. An illustration of an adaptive kernel is given below. Notice how the kernel vary in width depending on point density. The more sparse the data the wider the kernel.

The first step to running a GWR regression is to specify the kernel and bandwidth. The kernel is used in deriving the local weighted least squares estimate.

GWR Model for Land Prices in Brooklyn

The process of determining whether GWR is required is to (1) get the best fitting global equation and determine if spatial heterogeneity exists, then (2) select a kernel type and its bandwidth size, (3) run a GWR on the best fitting global equation, and finally (4) test whether the local coefficients provide a significant improvement over its global counterpart.
The best fitting global equation is the SLM equation in the previous section where there is no significant spatial dependence yet heteroscedasticity exists. To select between kernels, we run a GWR with the adaptive Gaussian and adaptive bi-square options and compared model fit. From *Table 5.7*, the adaptive bi-square kernel option has a higher adjusted R-squared, and lower AIC fit.

Table 5.7: GWR Adaptive Gaussian versus Adaptive Bi-Square Kernel

DIAGNOSTIC TESTS	ADAPTIVE GAUSSIAN	ADAPTIVE BI-SQUARE
Model Fit		
Adjusted R-Square	64.92%	65.42%
Classic AIC	1311.44	1302.93
AICc	1316.57	1307.09
BIC	1468.67	1444.67

The extra complexity of varying regression coefficients is worthwhile only if the GWR model results in a smaller residual sum of squares in comparison to the OLS estimation. *Table 5.8* shows the difference in criterion statistic. A "DIFF Criterion," especially greater than or equal to two, suggests the local term is better treated as global. For our land price model, all the coefficients are significantly spatially varying so we maintain the GWR to the SLM model.

We enter the GWR parameter specification into GeoDa to replicate tests for heteroscedasticity and spatial dependence. *Table 5.9* shows the GWR regression has no significant spatial dependence and no significant spatial heterogeneity.

We summarize OLS, SLM, and GWR results in *Table 5.10*, including comparisons of model COD and PRD.

GWR Local Versus SLM Global

```
                         AICc Criterion
*****************************************************************************
 Geographical variability tests of local coefficients
*****************************************************************************
 Variable                  F                  DOF for F test  DIFF of Criterion
 -------------------- ------------------ ----------------- -----------------
 Intercept                 36.880106      2.342   532.304       -79.537595
 W6_LNPRC                  42.351908      2.164   532.304       -84.748390
 USEMAXFAR                  2.200540      2.061   532.304        -0.148811
 LNSIZE                    12.544813      2.724   532.304       -28.967626
 LNWF                     263.153674      2.147   532.304      -403.221354
 LNUNIV                    41.719456      1.597   532.304       -62.902446
 LNGAL                     29.591917      2.246   532.304       -61.307843
 LNRSTR                    15.363378      2.711   532.304       -36.399778
 -------------------- ------------------ ----------------- -----------------
 Note: positive value of diff-Criterion (AICc, AIC, BIC/MDL or CV) suggests
 no spatial variability in terms of model selection criteria.
 F test: in case of no spatial variability, the F statistics follows the
 F distribution of DOF for F test.
*****************************************************************************
```

Negative Values Favor Local Model

Table 5.8: GWR Local Versus SLM Global

GWR Tests for Heteroscedasticity and Spatial Dependence

```
DIAGNOSTICS FOR HETEROSKEDASTICITY
RANDOM COEFFICIENTS
TEST                    DF            VALUE             PROB
Breusch-Pagan test       1            1.6502            0.19893
Koenker-Bassett test     1            0.9536            0.32879
SPECIFICATION ROBUST TEST
TEST                    DF            VALUE             PROB
White                    2            1.0468            0.59252
-------------------------------------------------------------------

DIAGNOSTICS FOR SPATIAL DEPENDENCE
FOR WEIGHT MATRIX : W6_3G
   (row-standardized weights)
TEST                           MI/DF          VALUE          PROB
Moran's I (error)             -0.0335        -1.3778        0.16828
Lagrange Multiplier (lag)         1           1.3933        0.23785
Robust LM (lag)                   1           0.2315        0.63044
Lagrange Multiplier (error)       1           1.9929        0.15804
Robust LM (error)                 1           0.8310        0.36197
Lagrange Multiplier (SARMA)       2           2.2244        0.32884
```

No Heteroscedasticity

No Spatial Dependence

Table 5.9: GWR Tests for Heteroscedasticity and Spatial Dependence

DIAGNSTIC TESTS	OLS	SLM	GWR
Model Fit			
Adjusted R-Square	60%	63%	65%
AIC	1369.4	1330.36	1302.93
Schwarz Criterion	1426.75	1391.05	
Median Ratio (LNPRICE)	1.00353	1.00478	0.99466
COD (LNPRICE)	0.04785	0.04557	0.0400
PRD (LNPRICE)	1.0001	1	1.004
Spacial Dependence	Yes	No	No
Heteroscedasticity	Yes	Yes	No

Table 5.10: Summary Statistics for OLS, SLM, GWR

GWR performs well on all model of fit measures. However, the added complexity of GWR and the interpretation of local coefficients make it difficult to apply in practice. This is the case because it is difficult to explain locally varying coefficients to the public or to justify consistency of assessment when the model generating it varies point by point.
In practice, we take GWR as an explanatory tool to indicate additional steps that must be taken to address heterogeneity in the SLM model. For example, one can map the varying coefficients, form clusters, and use those clusters as area dummies in the SLM equation.

CONCLUSION

This section explained steps to diagnose and model spatial affects in CAMA models using data from land sales in Brooklyn, NY. Using a well-known open-source software called GeoDa, we explained the diagnostic steps to determine if spatial autocorrelation exists and if so, whether SLM, SEM or SARMA is the appropriate model specification. Using another well-known software called GWR4 which was developed to address local regressions, we showed how spatial heterogeneity can be tested and modeled.

Prediction to unsold samples is possible. For SLM/SEM/SARMA models this requires the creation of the weight matrix for any point in the area, whether sold or unsold. The SLM estimated equation can then be used for the weighted spatial lag formed for the unsold parcels and their physical characteristics.

Similarly, GWR4 has an option to predict local coefficients by uploading (x, y) coordinates for unsold parcels. In practice, GWR should be used as an explanatory tool and not as a final CAMA model. The spatially varying coefficients can be used to determine cluster areas. The final SCAMA model would include cluster areas in the SLM equation as explanatory variables.

CHAPTER 6
Using GIS to Serve the Taxpayer

1 | Introduction

Most assessors love to explore maps, and mapping data can be particularly helpful to both the public and government officials. This chapter focuses on utilizing the capabilities of GIS to visually present data to these two bodies. Since the expectation that government records be accessible to the public has become commonplace, the use of GIS for property tax record display has also become more and more prevalent across the world. New GIS tools provide property owners and concerned citizens with easy-to-use search instructions that allow them to sort and filter through property tax records. Public access to property tax records in a GIS leads to greatly improved taxpayer interaction and feedback. Using GIS tools is currently the most dynamic and cost-effective way for every office involved in land and property tax administration to communicate valuation information in a visual format.

The capabilities endowed by a GIS tool greatly facilitate individual research and mapping discovery, and are therefore a huge asset in the administration of property tax systems.

2 | Government Reporting for Valuation Offices

2.1 | Historical Practices

For valuation offices, government reporting has historically been accomplished by transferring large data sets to oversight offices or by providing them with tabular spreadsheets. Information given to the public, on the other hand, was reported through long lists printed in the newspapers—a practice which later advanced to the publishing of online tabular reports.

Many valuation offices now use online programs that allow taxpayers to search for valuation data and information relevant to the development of property tax values for real estate. The search functionality of online property tax data allows taxpayers to find a specific property (by entering a property identification number, or PIN) and review its data with just a few clicks of the mouse. The public should always have instant access to individual property lookup, and using GIS will bring property tax record reporting to a whole new level.

2.2 | Reporting Property Tax Records Online—Maps Really Matter!

GIS technologies bring a visual element to the examination of property tax records, which makes property tax administration a more accessible and rewarding experience for the public, government stakeholders, and policymakers. GIS tools give land and property tax administrators the chance to create the best public-facing toolkit possible to increase public understanding of property tax administration processes and valuation development.

With GIS, taxpayers have the ability to review properties' estimated values, and—importantly—to compare them to other, similar properties. Government transparency (including of these kinds of records) is a goal that we all share, and GIS mapping helps to clarify why a government has made certain decisions.

Valuation agencies that do not have a GIS platform (the software or hardware that house the GIS) to display records should explore their options for purchasing GIS interfaces. Electronic property record data sets can be easily imported into a GIS application if there is a geo-reference code (which there often is) within the existing property records. Inter-government collaboration—meaning, cooperation and resource sharing among government agencies—has catapulted the sharing of GIS data and GIS platforms providing even the smallest jurisdictions with a new and cost-effective avenue for developing public mapping displays and interactive taxpayer tools. Pursuing a collaborative government agreement to utilize an existing GIS platform is well worth the effort.

GIS tools offer taxpayers the ability to create their own maps, and to conduct research and analyze data on their own. GIS is designed to accommodate the compilation of comparative property tax data on a map. Since valuation-related overlay data often include sales, income, cost, or other relevant data characteristics, mapping allows simplified visualization. This means that one can easily see the exact property characteristics and data assigned to each property.

Providing the public with the newest GIS interactive tools—online mapping programs that allow users to create their own queries and therefore decide exactly what data are displayed and mapped—is the best option to pursue. Interactive GIS tools allow for data to be filtered quickly, and can also be used to analyze comparisons of properties and their underlying characteristics. Interactive GIS software will reduce taxpayer inquiries to your agency by allowing for individual research and investigation that produces immediate results.

Having public-facing maps and interactive mapping tools will **increase public understanding of your property tax administration processes and the rationale for your decisions.**

2.3 | Explaining Property Values

As mentioned, valuation offices often use maps created within a GIS platform to display the data that were used to create property tax value within their agency. These data often include:

- Recent sales data
- Land dimensions
- Building characteristics
- Income and cost data for various property types
- Locational indicators, such as a neighborhood code or locational influence factors
- Other external influence indicators
- A defined classification of the property type which was used to categorize the property for valuation development

Spatial display, or mapping, of the common indicators used to develop estimates of value for various property types allows taxpayers to sort and filter through data to find the most comparable properties—that is, properties whose characteristics are most similar to the property in question. Being able to review mapped information online gives the public an important understanding of how property values are estimated, so that they are able to discover, on their own, the uniformity, accuracy, and fairness inherent to a great majority of property value estimates.

▸ 2.4 | Finding the Most Comparable Comparables

When explaining how property values were developed, important factors that must be highlighted are the locational influences on a specific property. Contrasting one property to another property with different locational influences will underscore the impact that location has on value and property comparability.

Public access to mapping tools has eliminated the hours of effort that used to be requiredto find tax records for the most comparable properties. Gone are the days when taxpayers must wade through numeric property number listings of tabular data—now, a property's location factors and other descriptive data are displayed on visually appealing maps. It's clear that GIS is a much better tool for searching and illustrating property tax record comparisons.

▸ 2.5 | Mapping Property Value Change

It is often helpful to display rates of property value changes both by location and over time. Mapping yearly comparisons of property values can assist taxpayers in tracking increases or decreases in a property's value over a set timeline. Taxpayers can also compare their own property's record against the value changes that have occurred in similar properties in their area.

Taking it one step further, by displaying new construction or recent permit information, taxpayers can better understand the variances in property value increases. If one property in an area experiences a larger increase in property value than other similar properties, the explanation for this higher percentage can be often be found through additional research. Information regarding renovations or new construction can help determine why that property is now more valuable or experienced a higher value increase than comparable properties in the area.

▸ 2.6 | Displaying Information to Assist the Public in Property Valuation Appeal Decisions

In this industry, it's unavoidable that eventually some taxpayers will be unhappy with the value their property is assigned, and they might want to lodge an appeal. Here, maps that show both property value and characteristics data come in handy. Since location is a prime determinant for choosing comparable properties, having a way for taxpayers to actually see those comparable properties, and the variables that determine their value, is an incredibly useful tool for justifying a property's assigned value.

If a property owner does decide to appeal their current valuation estimate, relevant data and information displayed on the public map can help the taxpayer prepare for the appeal process. A valuation office will be able to display the number or location of property appeals that were filed within that jurisdiction, and sort these appeals by different types of property classifications. The result of each appeal can also be recorded to give further information on the overall uniformity of property values. In this way, mapping property values and data to describe and characterize comparable properties can help a valuation agency defend and further explain values to taxpayers and oversight agencies.

These easy-to-use GIS mapping tools are the best way to visually explain how a property's value was developed. This helps many different stakeholders: the public, other agencies tasked with higher levels of valuation appeals or valuation oversight, and the assessment office itself. Data can be deciphered more easily when an office is able to visually display uniformity and equity trends or patterns across any government jurisdiction. This is an unparalleled advancement in the level of transparency offered for property tax administration. For these reasons, the use of GIS to exchange property tax record information will become extremely common—if not expected—in the future.

Figure 6.1: Example of Map with Property Tax Data. *Copyright/Source:* Esri and its licensors

3 | Mapping to Assist in the Administration of Valuation Office Duties

GIS maps can be created to display other information that valuation officials routinely process and administer. Displaying the additional information that is retained for every property allows the public to easily examine and verify current property record information.

3.1 | Data Verification

First and foremost, the public and taxpayers want information about what is contained in each property tax record, which can be easily displayed if property tax record information is integrated as a data layer within the GIS. Property tax records can be mapped by taxable year to allow for the easy identification of PINs and the verification of both current information and historical recordings. Having Property Identification Numbers (PINs) mapped alongside property location addresses allows taxpayers to confirm that they are looking at the correct property, and adding multiple years of property tax records allows for even further historical verification of any property record information.

This is important because the PIN assigned to an individual property might change over time. There are many different reasons that a PIN might change, but some of the most common are: (1) a division of one property into multiple properties, (2) a consolidation of two or more properties into one single property, and (3) changes to the legal ownership of the property. Large condominium projects with multiple buildings and a phased development overtime, for example, can prompt many changes to the associated PINs each year. That is why it is so important to have measures in place to ensure that a property owner is investigating the correct PIN.

The Cook County Clerk (of Cook County, Illinois) partnered with Cook County's GIS department to create a GIS application that could coordinate the maintenance of property records and maps over time. This application—nicknamed PINMAP—is used to maintain annual property tax maps, along with land and property tax record information. The main menu of PINMAP is shown in *Figure 6.2*.

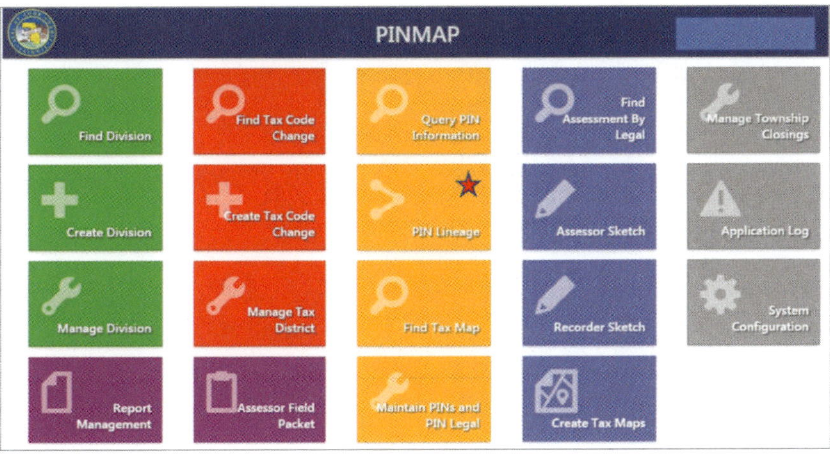

Figure 6.2: PINMAP

Each tile on the screen represents a different function within the PINMAP program. These functions assist in property record updating and operational processing, and include a property index (or identification) number (PIN) lineage tracking system. The PIN Lineage application (indicated by a red star in *Figure 6.2*) was developed to help the public keep track of rapidly changing PINs. The Clerk's office can enter a property's PIN for any tax year, and the correct property will be found, even if that property has been assigned a new PIN in the years since.

Figure 6.3 is an example of the simple visual display that shows the transformations that a PIN might go through over time.

Once they have selected the PIN that they want to review, the Clerk's office can investigate all of the other information contained, by year, within that property's tax record. Clicking on a specific PIN will immediately bring the user to the relevant documents from which those PINs were created.

Another reason that it is vitally important that taxpayers be able to instantly find

Additional functionalities of the PIN Lineage screen (listed in columns) include:

Parent PINs. This column displays any information regarding prior property index numbers that a specific PIN was known as in the past.

PIN. The center column shows all information that includes that specific PIN, followed by information on any related properties.

Child PINs. This column displays any information and PINs that were created from a specific PIN.

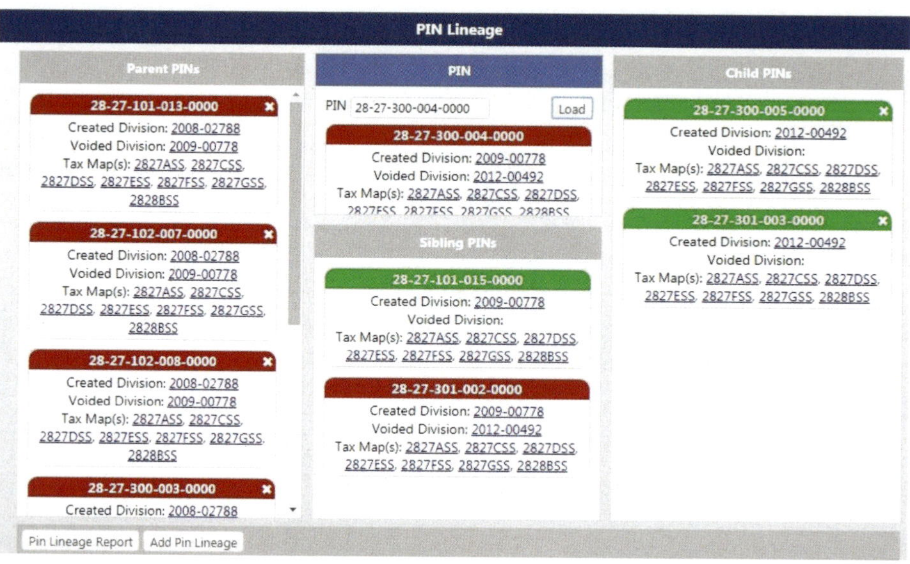

Figure 6.3: PIN Lineage

accurate information on their property tax records is that property taxes are often associated with a property's PIN, rather than its address. Taxpayers must be able to find the right information in order for taxes to be collected accurately.

▸ 3.2 | Exemption Verification

Another common function of many valuation offices is the administration of property-tax-related primary residence or homeowner exemption programs. GIS maps allow for the geovisualization of such programs' results.

For example, maps can be used to display which property owners have received a homeowner's exemption, a senior citizen exemption, or any other type of property tax exemption that is authorized for that jurisdiction. There are two important reasons for mapping property tax exemptions across a jurisdiction: (1) so that taxpayers can review exemption records for their property, and (2) such mapping will assist the valuation office in identifying underserved areas, which might not be informed about the opportunities they have to reduce the current taxable value of their property.

▸ 3.3 | Mapping Property Tax Districts (Including Incentive Districts)

Maps can identify whether a property resides within the boundaries of some type of economic development zone or district mandated for property tax collections. It is important for a taxpayer to know every taxing agency—including special zones or districts—that their property resides within so that they know how their property is ultimately affected by any additional property tax program.

The agency responsible for providing information on all taxing authorities which compose a property's property tax bill—if this is not already a function of the valuation office—should display the boundaries of every taxing authority within a data layer of the GIS. The taxpayer can then conduct further research to determine the amount of property tax dollars due to each taxing authority, based on the physical location of their property.

▸ 3.4 | Enhance Taxpayer Online Interaction and E-Government Tools in Administrative Duties

Virtually all GIS platforms allow for the development of online interactivity and add-on applications. Interactive email systems let taxpayers communicate with assessors and provide a record of those interactions, which makes it easier to track taxpayer requests, such as changes to a property record.

Although many valuation offices already have interactive components for taxpayer services on their website, offices can now get very creative with the interactive processes available through GIS. Taxpayer inquiries can be alleviated, for example, by offering maps that display valuation office information and that allow for information searches to be conducted.

GIS can enhance taxpayers' ability to interact quickly and easily with their government. They can find answers to their questions and increase communication by using the following online functions:

- **Requests to update property records** – Property owners should be given an opportunity to review their own property record online, and should be able to send an official request through this system that the government office update any errors they notice in the recorded property data characteristics.

- **Q&A with the valuation office** – Property owners and taxpayers can submit questions directly to the valuation office. Mapping can augment this process further.

- **Taxpayer exemption applications,** with taxpayer or property owner verification keys.

- **Taxpayer appeal processes,** including mapping displays of comparable properties.

- **Information requests for maps or government data** to be delivered for spatial formatting and display.

- Information requests per **public information laws and statutes.**

- Information requests from **local media sources.**

- **Requests for an office representative** to be a public speaker for an upcoming event.

GIS advances online operational efficiencies by reducing paper processing and offers more upfront information for taxpayers. Online tools provide convenient and efficient **e-government services** for taxpayers, particularly when it comes to functions related to property tax appeal and exemption applications. Some taxpayers might not be able to submit applications in person, and (if laws allow it) program applications can be offered through an online process in addition to being sent through the mail.

Augmented taxpayer interaction tools can reduce the number of phone inquiries and the amount of foot traffic that comes through your office. The public now expects to have instance access to information and to e-government services, and GIS allows valuation offices to meet that desire.

3.5 | Inter-Governmental Cooperation and Data Sharing

Since the data collected by a valuation office are always land- and property-ownership-based, the easiest way to display and share these data with other government offices is through the use of an electronic mapping system. As mentioned earlier, sharing such data with entities that are tasked with a higher level of appeal function, or with administrative oversight of valuation processes, will allow those agencies to geovisualize current property values. Sharing critical data with other government agencies is cost-efficient and increases transparency across the board. It is a sensible practice that is fast becoming an expectation.

Authoritative data compiled by valuation offices are **invaluable to countless agencies at every level of government.**

Authoritative data refer to data that are considered to be reliable and accurate. An authoritative data source is the office that compiled the data as part of its mission or function. Authoritative data compiled by multiple valuation offices within land and property tax administration often include:

- Legal land boundary definition and ownership information (which usually comes from the register's office)
- Land dimensions, building data characteristics, and value definition (from the valuer's office)
- Identification of property tax districts, district taxing rates, and record of payments (derived from the clerk or treasurer)

These data are invaluable to countless other government agencies. The most critical example is that agencies at all levels of government worldwide who need to respond to emergency situations and natural disasters rely on accurate property ownership and property data characteristic information. Emergency response services need a clear understanding of the current infrastructure of individual properties. Property records from local property tax administrators often provide a foundation for emergency responders to work from.

Other agencies that have responsibilities related to real estate records also rely on the valuation office's record. These include:

- Agencies responsible for issuing permits related to property within the jurisdiction. Such agencies need to verify property types or property classification through accurate data and descriptions, and this information can assist in accurate permit issuance.
- Agencies responsible for the development of land and building zoning laws, and their designated boundaries. These agencies use valuation office data to conduct impact analyses of any zoning proposals.
- Agencies responsible for creating proposals for economic development related to property taxes, including new urban housing and "smart" cities. These agencies benefit from accurate descriptions of current property records, and are imperative for government planning.
- Government units responsible for the administration of delinquent property taxes also need accurate property records to administer the duties of delinquent property tax sales or property tax redemption procedures.

- Transportation and public infrastructure authorities will often ask for assistance in conducting public transit needs analyses in order to verify current property data and values.

- Legislators and government bodies. These groups need property data and information to measure the costs and benefits of proposed tax policy solutions. It is not uncommon for elected officials to ask the valuation office to conduct data analyses, or, at the very least, for the office to express an informed opinion on a proposed property tax policy law.

The list of other public service and government agencies—from the local level all the way up to the national and international level—could go on and on. Future efforts to share and standardize data between these groups will be discussed in Chapter 7.

3.6 | Visualizing Policy Decisions and Program Development

Many GIS applications allow the public to download data and information contained within a GIS platform for their individual analysis and review. Access to government data through GIS has been extremely beneficial for academic communities and public policy stakeholders, and has led to the promotion of good governance in the development of property tax policy.

A government valuation office has enormous amounts of authoritative data that can assist in public policy development and implementation, particularly within the field of property tax policy.

The data taken from a valuation office are an incredibly useful starting point for forecasting and projecting anticipated fiscal impacts of the proposed program. Whenever there is a new property tax program, there could be "winners" and "losers"—property owners who now pay less in taxes than they did before, and property owners who now must pay even more. Part of the analysis that is conducted includes investigating specific property types in order to identify the affected parties.

Through GIS, the findings of this research can be mapped visually, rather than simply presented on spreadsheets. As they say: a picture is worth a thousand words. Tax policy programs can be mapped with other data as well, including property values, in order to visualize a program's causes and effects on properties that ranges from the very lowest to the very highest in value. The impacts of a program can be mapped and discussed in terms of fairness and equity issues, too.

Many initiatives have been analyzed using authoritative data from a valuation office, including:

- Targeted property tax incentive programs
- Targeted economic growth programs
- Exemption programs for homeowners, seniors, long-term homeowners, veterans, and people with disabilities
- Other initiatives

Before a legislative or other governing body approves a new property tax policy program, research is often conducted to assess the fiscal impacts on the overall **property tax base** (the total property tax dollars collected, along with any potential shifts in tax dollars that are collected for specific tax districts), including the estimated impact on property tax rates. In order to conduct this extensive research, agencies within the property tax administration process often work together to share their resources and data.

Different agencies whose resources can be drawn from include the recorder or property registry office, the valuation office, the tax rate calculation or tax extension office, and the treasurer or property tax collector office. Together, these agencies have an enormous amount of authoritative data that can be used for valuation and for the development of tax policy.

Cook County's Tax Increment Finance (TIF) Viewer is a prime example of intergovernmental cooperation. The TIF Viewer uses collaborative property tax record information from multiple property tax administration agencies and GIS mapping tools to increase tax policy accountability.

Over the past 30 years, tax increment financing (TIF) districts were created and enacted by legislation throughout the United States in order to spur economic development in specifically defined boundaries and qualified areas within jurisdictions. Cook County, Illinois, is one of the largest counties in the United States and has a significant number of TIF districts within its boundaries. The Cook County Clerk asked Cook County's GIS department to come up with a creative solution for viewing these districts. Together, they developed a TIF Viewer that utilizes GIS software.

The calculation of property tax dollars going directly to TIF districts is somewhat complicated. Basically, though, after the creation of a TIF district, any amount of annual growth in property value that generates more property tax dollars from the TIF's inception year is directed specifically to the TIF district. Since property tax dollars are now going directly to the established TIF district, the TIF is essentially diverting these dollars from other, underlying taxing districts (e.g., schools, public libraries, water reclamation districts).

Taxpayers can analyze all of the established TIF districts and their boundaries within Cook County. The Cook County TIF Viewer also displays:

- All properties and relevant property value information for each property within the boundaries of the TIF district
- The total equalized property value of all properties within the boundaries of the TIF district
- The total amount of property tax dollars, collected every year, that go directly to each established TIF district

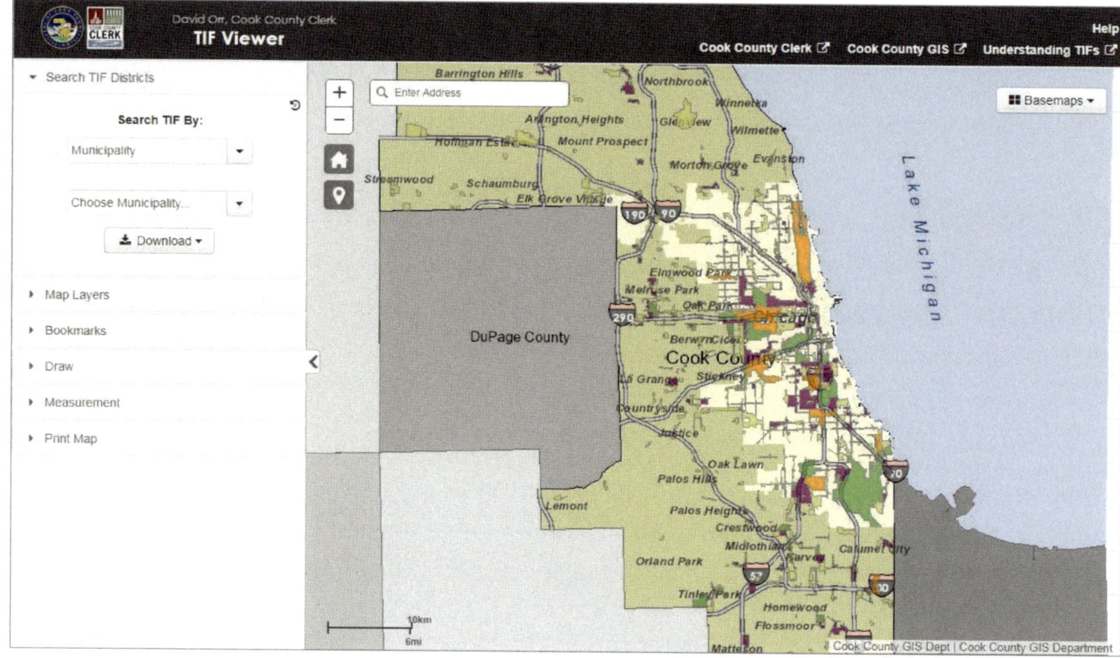

Figure 6.4: TIF Viewer, Main Screen

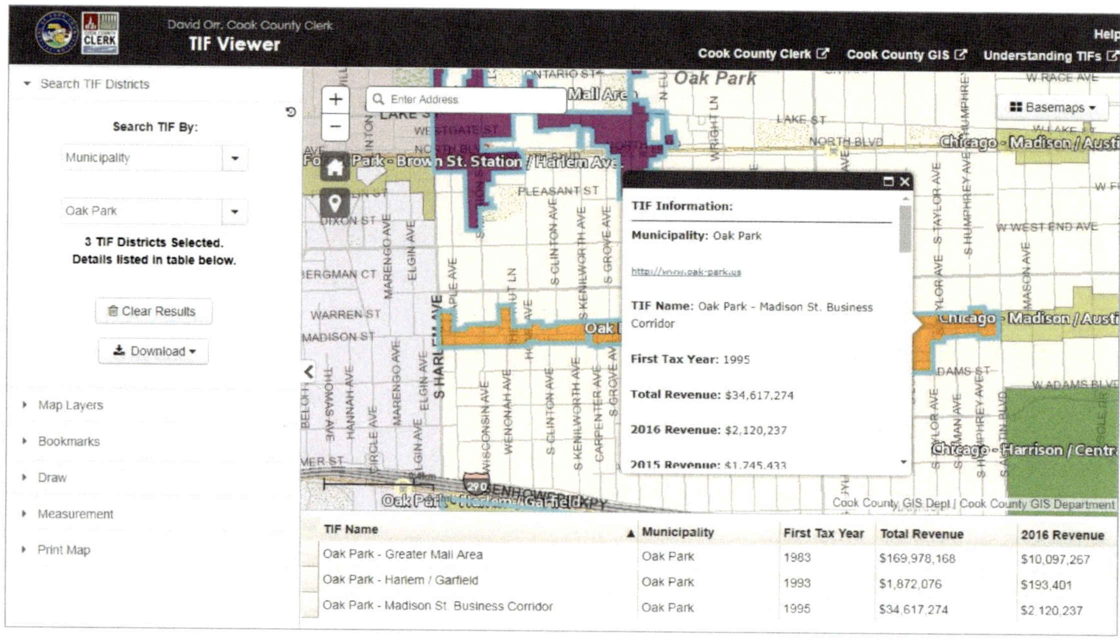

Figure 6.5: TIF Viewer, Sample Information

The TIF Viewer was created to understand the amount of property tax dollars that had been diverted into TIF districts over the 17+ years they can exist. The designated boundaries for every TIF can now be easily viewed on a map, so a taxpayer will know if they own a property that sits within a TIF district. *Figure 6.4* shows the main screen of the TIF Viewer.

All TIF information housed in the Cook County TIF Viewer, including the annual amount of property taxes collected for each district, can be downloaded and analyzed by individual researchers and policy analysts. The easiest way for the Cook County Clerk to describe and display this information was to map it using this mobile application. *Figure 6.5* shows an example of some of the information that can be displayed.

This information is essential for reviewing the impacts of the TIF tax policy program. In addition, itemized information on individual Cook County property tax bills informs a taxpayer whether or not they reside within a TIF district, and the exact amount of money from their property tax payment that is being paid directly to the TIF district.

Taxpayers, government administrators, and lawmakers need to understand the fiscal impacts of existing and proposed property tax programs. Tax policy analysts and experts need GIS tools like Cook County's TIF Viewer so that they have instant online access to critical data related to property tax, and can easily visualize the effects of a tax policy decision and review a program's performance. This TIF viewer is a great example of how GIS can be used to maximize intergovernmental cooperation and improve government transparency.

▸ 3.7 | Taming Public Open Data Portals with GIS

Many government agencies have established **open data portals**—websites used as a repository for multiple data sets, which often house a huge amount of government data and information. Some portals have extremely advanced search engines, while others simply amass data sets without much added search functionality.

Certainly, many government portals are improving over time by advancing their search capabilities in terms of the amount of data that can be displayed. Using maps and GIS tools to interact with open data portals makes it easier (and more fun!) for taxpayers to find and interpret information.

Open data portals that interact with government GIS (meaning that the open data are embedded as GIS data layers) allow for quicker identification of data and easier-to-use data filtering functions. Once users have chosen the specific data they want to analyze, they can view this information on a map. Many users conduct further iterations of mapped data, as they continue to analyze and discover new patterns of displayed information derived from their government's open data portal.

GIS provides the construct to make open data portal data much easier to use, and allow for visual display of information chosen to be mapped for individual review and analysis.

Since the GIS software platforms used all across the world are still fairly similar, having this information embedded in GIS also facilitates a vast exchange of information among government agencies. A common goal and continued goal for all valuation offices worldwide is to improve how these offices provide information to open data portal websites, and as part of government GIS. In Chapter 7, we will discuss future data exchange and standardization efforts.

3.8 | Data Sharing and Property Record Compilation

GIS mapping can display all property-related valuation information from various agencies, plus additional information from the government's open data portal, all at the same time. For example, a taxpayer's property can have information displayed that includes not only the characteristics of that property, but also what school districts (and other taxing districts) the property resides in, what politically defined boundaries the property resides in, and other property-specific government information.

GIS tools can create a visual display that gives taxpayers a complete picture of government information regarding a specific property. Essentially, GIS facilitates a composite government record that can easily be visualized on a map. This gives the public as much information as possible about all government agencies and districts that affect or interact with their property.

Secondly, all mapped information regarding an individual government property record can also have website links to that specific government agency that is responsible for the administration of that function. GIS functionality creates one-stop shopping for government information, research, and easy contact access, and makes huge strides in e-government service to the public. As mentioned earlier, this type of access to government records and services will be a basic expectation in the future.

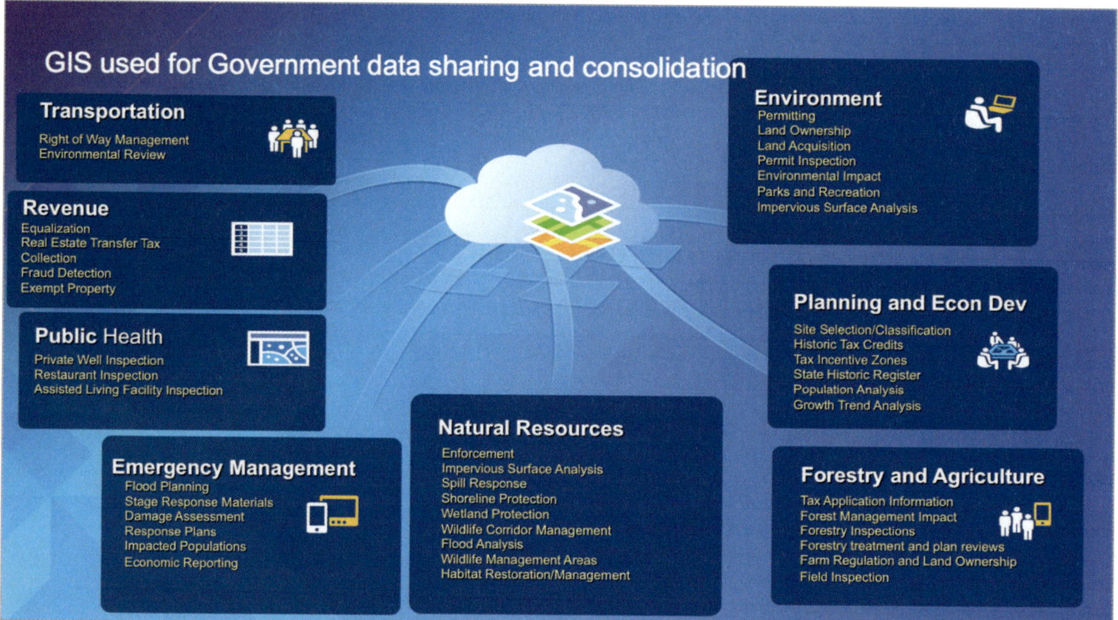

Figure 6.6: Shared information in GIS. *Source:* Brent Jones, Esri

3.9 | The Emergence of Integrated Property Tax Systems

Over the past 30 years, many property tax systems have been automated. Computer applications were developed for a single office or department within the property tax administrative process, but there wasn't much sharing of information. For example, the registry of deeds office would automate the recording of land records, which included information that defines the land dimensions and ownership of that property. This information would then reside solely within the land registry office. The property valuation office would then have to request information and records from the land registry office, either by asking for permission to access the data online, or by asking that annual data updates be given to the valuer's office so that they could conduct the initial stages of property tax valuation development and update their annual tax records.

The property valuation office would also have their *own* database (often housed on a different software platform) which included property data characteristics. Meanwhile, the clerk and the treasurer—or other offices primarily responsible for the extension and calculation of property tax rates, the preparation of property tax bills, and the collection of property tax dollars—would have their own individual computer systems as well. Needless to say, this was not an efficient structure.

Segmentation within property tax administration **resulted in strains on work efficiency and barriers to data sharing.**

Over the past two decades, GIS has given us the ability to map and analyze data by adding and overlaying separate data sets and data layers from *all* participating property tax administration offices. Information from various offices can now be integrated into a single mapping display.

An **integrated property tax system** can be defined as a software solution that replaces the multitude of unconnected software systems. These integrated systems unify multiple databases that are housed within different valuation offices, and allow users from any office to access any of these databases' property tax-related applications.

With an integrated property tax software solution, **the same software will be shared by all government agencies involved** in the administration of land administration and property tax systems.

Having a single software solution that combines multiple offices' databases facilitates the easy access of information by any office, and reduces unnecessary duplication of effort across government agencies. An integrated system can perform all of the operations and functions related to property tax administration, which again provides the value of unity. Modern GIS computer platforms and applications are compatible with many different integrated property tax software systems.

As we discussed in Chapter 2, GIS is **the best medium for maintaining, updating, and making changes to information contained within property land records.** There is no other system that provides such accurate and cost-efficient data management capabilities.

Importing data from other integrated systems into the GIS software also allows for all aspects of property tax administration to be displayed visually (e.g., through spatial displays and geovisualization), which makes it easy to share this information with the public. See Chapter 5 for a description of how GIS spatial analysis and mapping can further refine and advance mass appraisal models by adding more variables related to how a property's location influences its value.

GIS tools increase the efficiency with which valuation offices operate, which facilitates better performance. Having a single integrated software and GIS-mapping system that functions across multiple agencies and at all levels of government makes property tax administration run much more smoothly.

The processes and functions of tax administration that are made much more efficient through the use of GIS and integrated systems include:

- The recording and maintenance of accurate property records over time
- All operational processes involved in property tax valuation
- The calculation of property tax rates (commonly referred to as the tax extension processes)
- The ultimate creation of new property tax bills

While new GIS platforms are able to easily integrate with other mass appraisal software systems, it might take a bit of work to integrate older GIS platforms. These older platforms can still be programmed to import and export data among the two systems. Even so, the most cost-effective option might be to adopt an entirely new web-enabled GIS platform, and option that should be investigated as an alternative to integration.

3.9.1 | Challenges in Integration

One of the biggest challenges in creating an integrated property tax system is analyzing all the segmented data sets that currently exist within different agencies that administer property taxes. It takes quite a bit of brainstorming and collective planning to take the existing data sets—in their current, segmented structure—and envision the processes that new, integrated data tables will be able to perform. It's useful to collaborate with IT experts to take full advantage of the possibilities that could be born of integrating these data properly.

A plan should be made that lists the steps needed to fully integrate the existing GIS with the new integrated property tax system. Once data systems are integrated, duplicative operational processes are often discovered that can be eliminated, which makes the overall process much more efficient. Integrated systems also allow any one agency to quickly access information from any other agency, which further increases their operational efficiency and optimizes intergovernmental communication with taxpayers.

Integrating systems requires a significant shift in the technology that agencies are familiar with, and so it is critical to have a plan in place to address any confusion or learning curves that are likely to arise once the systems are integrated. Take the time to educate your agency on emerging technologies so that they feel comfortable using them. This will ensure that the users get the most out of these technologies that they possibly can, and will maximize the cost return of any innovative technologies you choose to implement.

Creating a plan for the future integration of property tax systems takes time, and requires excellent communication between government agencies. Each agency should designate one or more people to lead this initiative so that it doesn't get lost amidst the agency's other work.

For more information on the steps involved in creating an integrated property tax system—and for advice on how to make sure this integration is successful—see Chapter 1, which outlines what is involved in creating interactive GIS applications. The processes are very similar.

3.10 | Web-Based GIS Applications

Advanced GIS platforms are now using **web-based** or **cloud-based applications**, meaning that GIS software applications and services are housed on a shared computing platform located outside of your organization, and can be accessed from anywhere. This allows for even greater opportunities for valuation offices to expand their current data sets. **Cloud-based GIS applications** allow for the interaction and use of large open data sets that are also housed within cloud-based computing environments. GIS cloud-based mapping tools, along with the additional application of other authoritative information found within government open data sets, has opened the world to a new, **augmented reality** for valuation development. That is, new information has been combined with existing data to provide a new perspective on an old concept. New housing statistics, economic trend data, demographic data, and other characteristics that could affect value are more likely than ever to be housed within an open data set.

Web-based GIS also allow users to sort and filter through these open data sets. Users can import data relevant to their agencies into their GIS and filter out the rest. As additional open data sources are discovered and incorporated into a GIS, valuation analysts can analyze these data and determine whether they can be used to augment the current determinants of value within that office.

As discussed in Chapter 5, new value influence factors—such as data regarding traffic patterns, building views, economic forecasts, and other macroeconomic considerations—can also be added to a model to enhance valuation estimates. For example, discovering (through the application of information found in the government's open data set) that a specific commercial corner is very active and has high gross sales receipts indicates that this is a high-value commercial market. This knowledge can assist in delineating a highly desirable area for commercial properties.

Locational influences are often determined by many external factors, outside of individual property information and data characteristics. Open data sources provide a huge amount of readily available information that can help develop and create new locational and external variables (such as economic and demographic influences) in mass appraisal modeling for all property types across the world.

On the other hand, this increased access to mountains of data has created a new challenge for valuation officials: determining *what* information and *how much* additional data are necessary to help in the valuation process. Data discovery, as this sifting process is called, is a skill that takes time to learn and requires continual refinement. Luckily, web-based GIS tools offer ways to effectively perform discovery as well.

Access to open data makes it much more challenging to experiment with new mass appraisal models. Government personnel who are thinking about incorporating open data sets into their current tool kit would be well served by reaching out to assessment colleagues who have been using open data sets for a while, and asking them what data they have found that create new influential variables in property valuation. Those that do will find that the sky is the limit!

See Case Study 6.1 ("Web-Based GIS for Assessment Professionals: Warren County, Missouri") at the end of this chapter for a demonstration of how these applications might be used in the real world.

3.10.1 | GIS Story Maps and Dashboards

Avid GIS users have started talking recently about the ability of web-enabled GIS to create personalized **story maps**. These allow the creator to upload additional information—including pictures, text, or even video—with geovisualized data, resulting in an interactive, informative end-user experience (*Figure 6.7*).

A GIS story map is, essentially, a map that tells your personally designed story. For the valuation office, this could mean a map that displays information from other government sources (including economic indicators, housing statistics, and demographics) to create a "future trends" map that illustrates the trends that other government experts have forecasted will affect your jurisdiction, and how these forecasted trends relate to current real estate value trends. Value data for one jurisdiction can be combined with other government data, and narrated with pictures and text, to describe your jurisdiction's "valuation story."

There is a wide range of valuation information that can be conveyed through story maps. Some other examples include:

- An analysis of trends over time, on a year-to-year basis (or alternate timetable, such as a revaluation cycle determined by jurisdictional law)
- Illustrations that teach the public about the property tax system
- Visual aids used to assist the public in completing valuation appeals or exemption applications (pictures and text can be added to create step-by-step instructions)
- Any other type of public outreach that would be enhanced through the use of a visual tool

Another new and useful tool is the **GIS dashboard**, which manages to display a multitude of informational maps and data trend graphs on a single screen (as seen in *Figure 6.8)*.

GIS dashboards are an excellent public-facing tool for explaining trends in valuation, trends in applications for exemptions, trends in the number of appeals being filed within a valuation office, or any other statistic that the valuation office has deemed important to share. GIS dashboards could also be used to visualize and provide answers to the public's most commonly asked questions. Sharing this kind of information through a GIS dash-

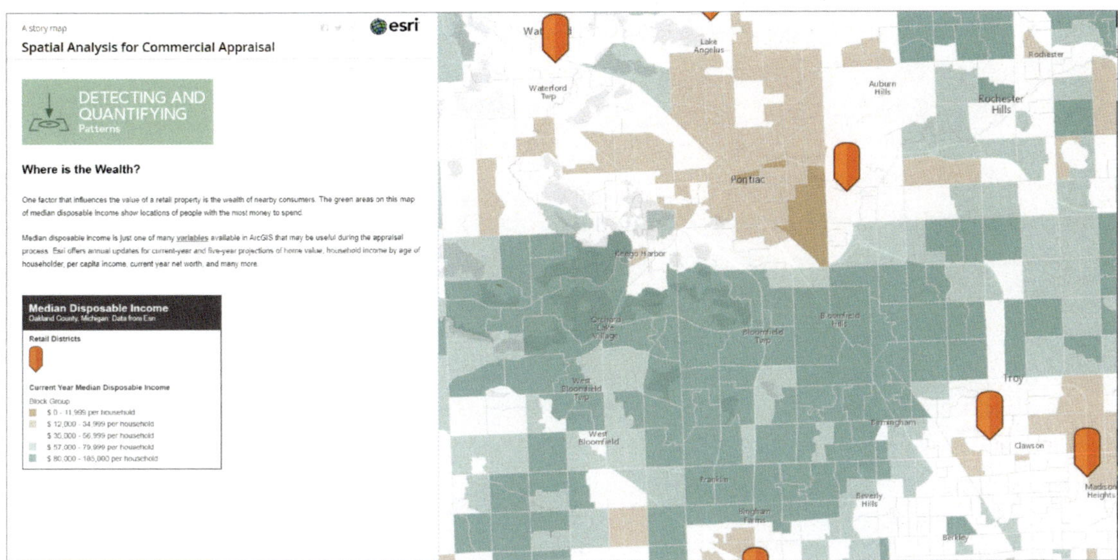

Figure 6.7: Story Map Example. *Copyright/Source:* Esri and its licensors

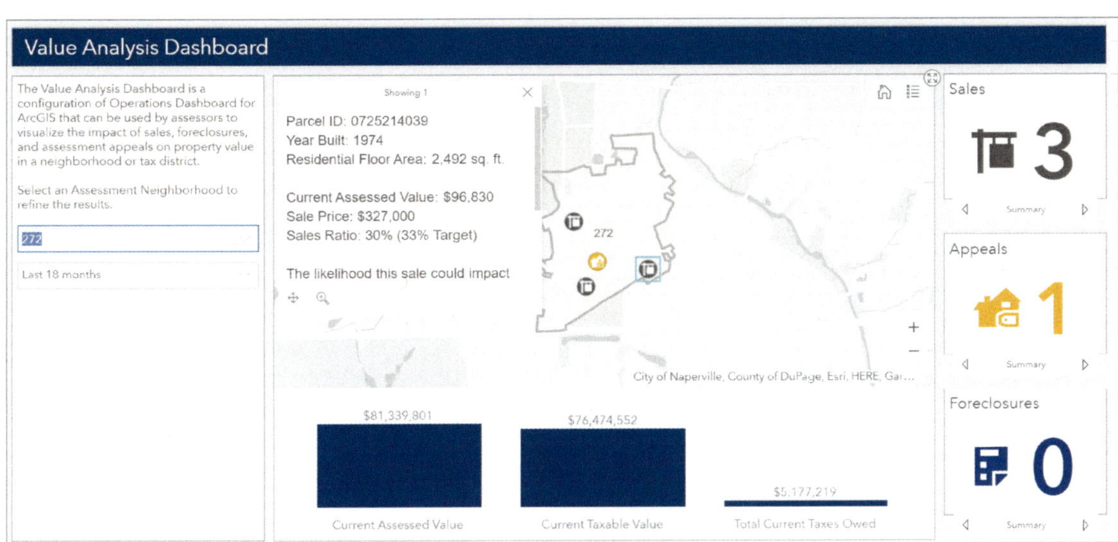

Figure 6.8: Value Analysis Dashboard. *Copyright/Source:* Esri and its licensors

board gives taxpayers a good starting point for obtaining a complete understanding of the tax system. Often, GIS dashboards also provide links to the various departments within the valuation office where people can find further assistance, if needed.

The use of GIS dashboards to report and display data will continue to expand over the next 20 years. The key to a successful dashboard that facilitates taxpayer services is to make it as interactive as possible.

4 | Conclusion

Since all operations within land and property tax administration are land-based, there is no greater tool than GIS for providing beautifully displayed information to the public. GIS provides taxpayers with an online source of information—a window through which they can explore government data. For further assistance and ideas regarding taxpayer engagement and interaction, check out the *IAAO Standard on Public Relations*. The provision of online assistance platforms—mainly through interactive GIS tools—is something that the public will expect to have access to in the future.

Case Study 6.1

Web-Based GIS for Assessment Professionals: Warren County, Missouri

Author: Russ Wetzel, GIS Analyst • Midland GIS Solutions

The impact of digital land data and the application of computer-based geographic information systems revolutionized how counties across the nation manage their lands. The advent and application of web-based GIS applications is now further transforming how counties continue to manage their lands, in an almost exponential way. County officials tasked with accurately assessing and mapping their land ownership boundaries, as well as tracking an increasing numbers of structures on each property, are turning more and more toward Internet-based solutions to increase efficiency in their land management processes.

As hardware and software technologies continue to develop, an increasing array of creative web-based GIS applications are likewise being developed, aimed at furthering the ability of counties to more accurately and efficiently manage their real data. Many counties are finding these online applications limited only by the imagination and the availability of quality data, and an increasing number of counties, across the country, currently utilize creative, custom-designed online applications to ease data access, enhance editing capabilities, and increase efficiency toward analyzing and reporting land data. These new web-based applications are fast becoming the new standard of daily land management operations.

Warren County, Missouri is a relatively small county in area, but is a progressive-thinking county in nature. The county is located in the eastern part of the state and is part of the St. Louis metropolitan statistical area. With an estimated U.S. Census population of 32,513, the county has maintained a steady annual population increase for nearly two decades. In 2016, they manage nearly 25,000 land parcels, including more than 2000 land transfers and roughly 100 land splits annually. County officials had witnessed the advantages of digital tax mapping programs early-on and had developed an in-house geographic information system relatively early in the GIS development movement. Like many early systems, however, inherent inaccuracies existed and, in their case, had rendered their particular system essentially unusable.

By 2006, Warren County Assessor Wendy Nordwald was looking for real solutions aimed towards streamlining their land management processes and the county partnered with Midland GIS Solutions, (MGIS), from Maryville, MO. Midland was already established as the largest GIS development firm in Missouri. Known not only for building consistent accuracy-based GIS programs, MGIS was also becoming known for creative web-based GIS applications. Warren County was particularly looking for something creative, and affordable, but also web-based, that would have a real impact on enhancing efficiency in their daily activities.

One of the first orders of business, however, involved a complete overhaul of the county's existing GIS data. Utilizing survey-based standards aimed at accuracy, and employing a host of available research, MGIS rebuilt the county's GIS from the ground up, including not only standard cadastral data layers such as parcels, lots, and roads, but also enhancing layers such as their 911 addresses, buildings on leased land, bridges, and soil layers, among many others.

Almost simultaneously, the county's web-GIS presence was being developed. This web presence was never intended to be limited to just a single web GIS, although MGIS did establish their custom web GIS program, Integrity™ GIS, for the county (*Figure 6.9*). This online GIS, within itself, was efficient, user-friendly, and immediately enhanced the county's ability to access, search, manage, and report their land data, but it also vastly increased the county's ability to share GIS data, not only between other county offices, but also with state and municipal entities, and with the public, in ways previously unimagined.

Warren County's developing web-based land management presence, however, went far beyond a single web GIS, to include numerous additional web-based applications. It was also not limited to a single company's products.

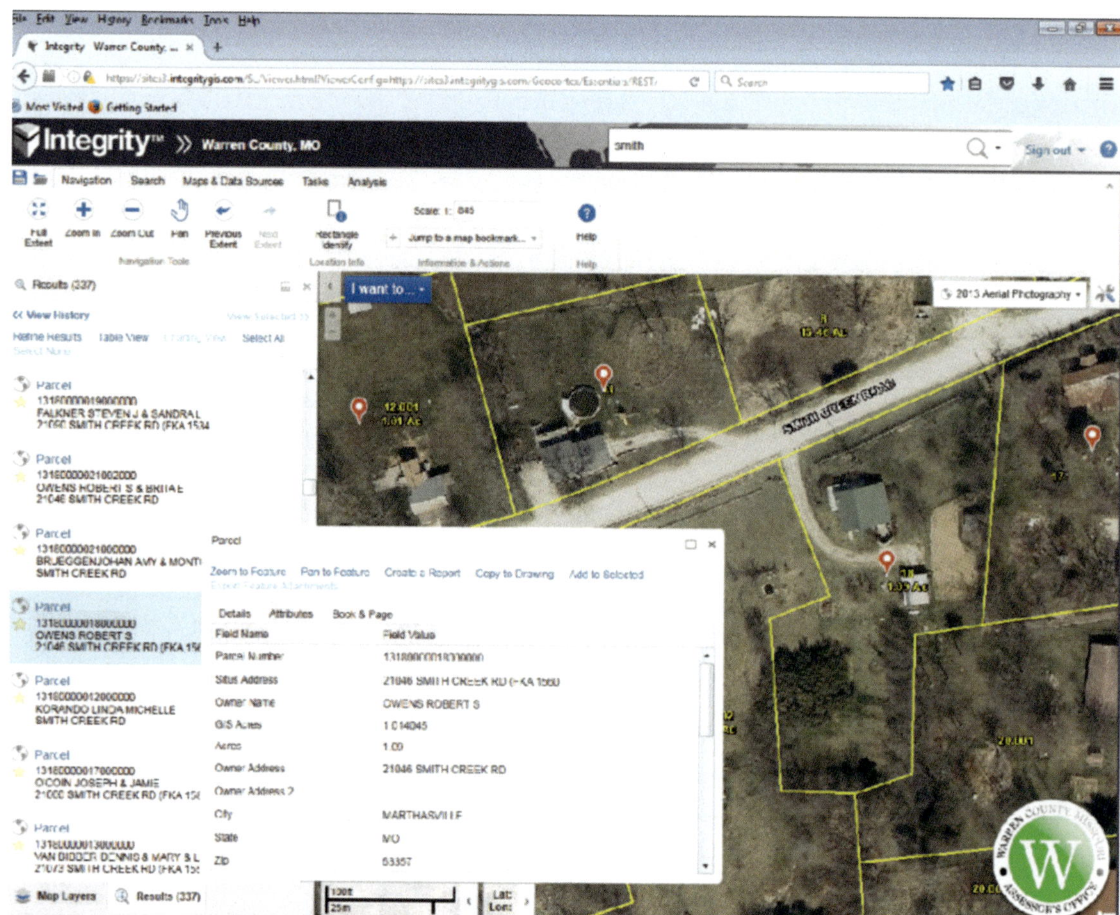

Figure 6.9: Integrity™ GIS Screenshot

LAREDO: ONLINE LAND RECORD MANAGEMENT

Developed by Fidlar Technologies, Laredo is a subscription-based online record management software utilized by Warren County to store and retrieve all of the county's recorded records, from affidavits to wills, from current deeds to historic research documents. Utilized by multiple county departments, the user-friendly interface allows personnel the ability to efficiently access records as needed, by record number, by date, by associated party, or by any of several other related aspects. For land management purposes, this online site not only allows various county personnel daily access as needed but also allows remote access. For example, Warren County does have their GIS land editing processes handled by a remote company. This online record retrieval site allows that editor access to the documents they need to successfully manage the county's GIS land changes.

PROPERTY SEARCH WEBSITE

From the county assessor's web page, MGIS set up a link to a basic, but customized, parcel search site, primarily developed for the general public to access and search for land information. A simple GUI offers the user a choice to search by PID, owner name, address, or value range. This allows users the ability to review a particular piece of land for a multitude of purposes, without requiring county personnel to take time to do this for them. Specific different security access levels were developed for different levels of users. For example, a subscription-based real estate company may have different data access from the free general public access. The Warren County assessor's office also provides another, similar, online CAMA database search site that allows users to search by account number, house number and street search, "doing business as" search, or name search.

PARCEL	NAME	ADDRESS	ASSESSED VALUE
05-10.0-2-00-087.024.000	BURNS PATRICIA L & SMITH PEGGY	2184 ANTHONY STEVEN CT	25751
04-16.0-2-00-004.001.000	FRALA VIRGINIA L & SMITH GORDO	101 FORT DODGE DR	16655
05-28.0-3-00-012.002.000	GOLDSMITH WESLEY W & STACIE L	4 SUNDANCE DR	20387
11-06.0-0-03-008.000.000	GUSDORF JEFFREY & SMITH LINDA	289 WHITE OAK POINT DR (FKA 2	18023
10-33.0-0-00-049.000.000	HOCKENSMITH JOSHUA C & KATHERINE R PARKER	20220 INVERNESS (FKA 220)	43168
11-20.0-0-00-001.014.000	JARVIS JUSTIN M & SMITH TAYLOR N	22485 S STRACK CHURCH RD (FKA	10124
05-26.0-0-00-003.005.000	KLINGINSMITH GREGG T & REBECCA L	16977 BOONE & CROCKETT CT	44819
04-16.0-2-00-004.028.000	MAHNESMITH ROY L & DENISE S	127 FORT HANCOCK CIRCLE	16498
11-08.0-0-01-001.000.000	MESSERSMITH ROBERT F & VIRGINIA A	227 CASTELENOVA DR (FKA 227 IN	16859
05-10.0-2-00-087.025.000	MOONSMITH JASON & CHELSEA	2188 ANTHONY STEVEN CT	20505
10-33.0-0-00-052.000.000	MORROW GARY W & SMITH BARBARA A	19533 COVENTRY CIRCLE (FKA 250	33182
11-16.0-0-00-022.771.000	NESMITH CHARLES L & VIRGINIA A	71 GENEVA BEND DR	10450
05-15.0-0-00-009.014.000	SCHWELLINGER MICHAEL R & SMITH	18491 WALNUT SPRINGS COURT	21041
05-31.0-0-00-013.000.000	SMITH ALLEN V & SANDRA	26381 HWY U	13771
11-10.0-0-00-026.026.000	SMITH ANDREW G	671 WHITETAIL CREEK DR (FKA 6	19086

Figure 6.10: Property Search Website Screenshot

WEB-BASED PROPERTY CHANGE FORM

This MGIS custom web page (*Figure 6.11*) was designed for interactivity between various land data management personnel, primarily CAMA data and map data editors. The simple interface allows personnel the ability to manage the same data, enter relevant parcel information as needed, and review and retrieve data as needed. It stores the data in a database format allowing ease of access for other online and in-house applications, eases research searches, and simplifies the process for generating data reports. The web site also provides a common location for specific issue questions and answers between personnel, without tying up time with telephone or email questions. It is very customizable, with access limited as needed.

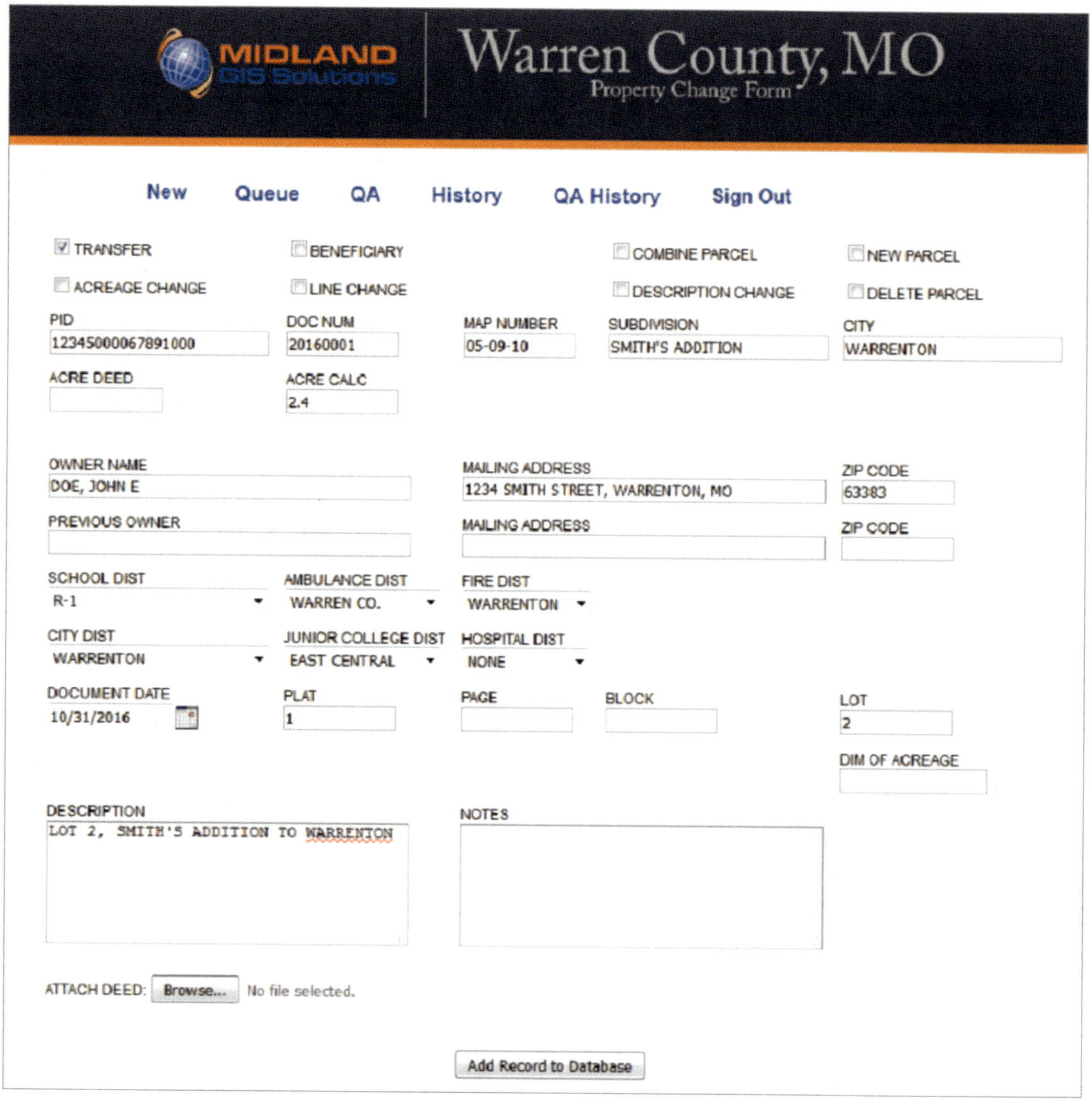

MIDLAND GIS Solutions

Warren County, MO
Property Change Form

New Queue QA History QA History Sign Out

☑ TRANSFER ☐ BENEFICIARY ☐ COMBINE PARCEL ☐ NEW PARCEL
☐ ACREAGE CHANGE ☐ LINE CHANGE ☐ DESCRIPTION CHANGE ☐ DELETE PARCEL

PID: 12345000067891000
DOC NUM: 20160001
MAP NUMBER: 05-09-10
SUBDIVISION: SMITH'S ADDITION
CITY: WARRENTON
ACRE DEED:
ACRE CALC: 2.4

OWNER NAME: DOE, JOHN E
MAILING ADDRESS: 1234 SMITH STREET, WARRENTON, MO
ZIP CODE: 63383
PREVIOUS OWNER:
MAILING ADDRESS:
ZIP CODE:

SCHOOL DIST: R-1
AMBULANCE DIST: WARREN CO.
FIRE DIST: WARRENTON
CITY DIST: WARRENTON
JUNIOR COLLEGE DIST: EAST CENTRAL
HOSPITAL DIST: NONE
DOCUMENT DATE: 10/31/2016
PLAT: 1
PAGE:
BLOCK:
LOT: 2
DIM OF ACREAGE:

DESCRIPTION: LOT 2, SMITH'S ADDITION TO WARRENTON
NOTES:

ATTACH DEED: Browse... No file selected.

Add Record to Database

Figure 6.11: Web-Based Property Change Form Screenshot

MOBILE WEB PLATFORM

This is a custom-built web GIS aimed specifically toward access through mobile devices. County appraisers access the map data in the field through a laptop computer, tablet, or smartphone with instant editing capabilities, greatly enhancing their ability to manage information related to new structures, or edit information related to an existing property or structure. Users simply click on an array of editing options to add or edit information. Likewise, they can take a picture of a given structure and immediately upload that image back to the in-house GIS and to managers. Active GPS, the latest available orthophotography, and up-to-date parcel and address information immediately assist the user to know exactly where they are, whose land surrounds them, what their contact information is, surrounding land values, or an array of other similar and pertinent information. Scale and labeling flexibility are designed into the application to create a very usable web-based format, greatly increasing the efficiency of this part of the process. Several additional and related web-based applications have also been created for differing applications, all aimed at streamlining the efficiency of the county land management processes.

One of the most immediate benefits of online land management applications lies in the ability to share pertinent data between other online entities, primarily other county departments, but also anyone needing access for whatever purposes. In Warren County, the sheriff's office routinely accesses the online data via the web, both for emergency and law enforcement response, but also for investigative and occupancy research, and for jurisdiction clarification, among other purposes. This includes data, both online and offline, available within individual police vehicles. Other departments, including planning and zoning, road and bridges, the county Clerk, and the county commission also routinely access land-related data via these web-based formats. Data sharing is not limited to only county departments. City of Warrenton routinely accesses zoning data, as well as the county's 911 addresses. State data requests, private entities, and especially the public, all gain greater access and efficiency accessing data they need throughout a given year. Another primary advantage of online data remains the ability to set access security, so that different given entities can only access what you need them to.

The web-based GIS applications employed by Warren County have greatly enhanced this county's ability to manage their GIS data. Online applications allow users from any location with Internet access the ability to search, locate, identify, assess, and generate reports, even from the convenience of any mobile device. The benefits of this level of access and the efficiency provided through web-based GIS access towards data management are constantly changing as more and more creative applications are being conceived and applied. The future of digital online land management remains exciting for those who require it. Web-based GIS applications will remain the primary focus for this access, moving ahead.

CHAPTER 7
The Future of GIS

1| Introduction

This chapter takes a quick look at the future of GIS in valuation offices in terms of the new technologies that will be available, the benefits that having open source software will bring, and the current efforts that are being made to standardize and share property tax records with a global community.

2 | Facing Current Challenges

One of the biggest challenges of maintaining GIS lies in ensuring that all users have access to the most current version of the software. Some users might be on older versions of GIS platforms and software, and consolidating these older platforms and preparing the office for conversion to a single, web-based, centralized application continues to be a challenge for many property tax administration divisions. Although upgrading these systems will be difficult, it will ultimately save cost, allow data sharing among agencies, and provide many more interactive tools to all users.

In order to gain support for an upgrade, tell stakeholders about the benefits that the additional capabilities available through cloud-based GIS will bring to routine government operations. A centralized, web-based GIS is the best way to meet the needs of all government agencies and provide an avenue for access to data, increased government transparency, and an easier incorporation of the most recent technologies. The data analytics that can be performed by any user are enormously useful, and teaching what these tools are, where to find open data sources, and how to create effective reports and maps for every department's needs are the keys to effective and lasting change.

The biggest challenge you are likely to face is how to collectively prioritize the GIS technology upgrades requested by property tax administration agencies. Decisions need to be made about the timing and priority of each and every request, with an emphasis on maintaining the accuracy of property tax data and developing interactive, public-facing mapping applications. This latter emphasis is important because giving citizens tools that help them understand government data helps to ensure accountability and gives the public a higher level of satisfaction with their government.

In order to properly prioritize upgrade requests, agencies will need to be able to communicate clearly and cooperate among themselves and with the governing body who authorizes the funding for all technology initiatives. Understanding the current challenges that impede the incorporation of new GIS technologies will provide a starting point for

Cloud-based GIS platforms are the best way to facilitate easy data exchange among government agencies, while the ability to geovisualize information and create user-friendly reports allows the public to digest complicated subject matter in a quick and easy format.

long-term planning. From there, action plans can be developed for implementing each new technology.

As open data sources expand, the sheer amount of data that will be available to valuation offices in the future will be enormous. The challenge will be in effectively applying this massive amount of data to local property tax valuation estimates. The value of real estate value will always be influenced by local factors, and therefore the proper application of macro data will be extremely important. Offices will need to be able to justify the choices they made to come up with value estimates. The best way to answer questions about how a value was reached is by mapping value estimates and including the reasons for coming up with those numbers, including the data indicators of value that were included in creating the estimate.

The advent and continuing growth of open data, while potentially daunting, should be viewed optimistically by assessors, as it can be used to improve valuations and quality control metrics. Open data published by environmental agencies regarding soil quality or other topographical information can be used to supplement land valuations and analysis. Census questions often ask households specifics regarding dwelling characteristics, which can provide additional insight for local markets. Population data collected through a census may also be paired with assessment data to offer new lenses for quality control, helping to identify, for example, whether or not areas of lower-income are being relatively over-taxed.

3 | The Future of Technology Advancements

As stated in Chapter 6, the integrated property tax computer systems of the future will be much easier to use, and GIS will become a vital and integral part of mass appraisal development, primarily due to the benefits derived from the visualization and geospatial analysis of data. GIS will facilitate the discovery of new mass appraisal location variables and an increased understanding of fine-tuned geographic influences on property valuation. As we move forward, GIS platforms will further integrate with other mass appraisal software—in fact, their current platform silos will most likely fuse together and become nearly seamless over the next 10 years.

GIS and other emerging technology breakthroughs are game changers for the entire world. With the invention of the Internet and cloud computing, the world now offers shared and interconnected hardware and software platforms that are external to each of our working spaces and house huge amounts of data and information, some of which is publicly available. Artificial intelligence (AI) programs created over the past few decades can teach computers how to "think" like the human brain in order to process information and make decisions. A new concept in machine learning, called **deep learning**, has also materialized in programs that dig deep to discover patterns and conduct data analysis using all of the open data and analytical resources available on the internet to unearth complex trends and patterns.

The term **internet of things** describes how smart electronic devices can exchange data online using sensors and digitally programmed components. Using these methods, these devices can be controlled remotely and report the information they learn to other devices.

Autonomous vehicles, sophisticated satellites, and smart robotics, and all other avenues of "smart" automation will dramatically change the form and function of valuation offices over the next 20 years. These so-called "disruptive" technologies have been embraced by GIS, which provides a bridge platform to integrate their use into all aspects of our lives.

GIS and new technology will streamline the operations of valuation offices in all areas, including the availability of electronic data; the collection, routing, verification, management, and analysis of these data; and how the office interacts with the public. The future of GIS will be determined by the imagination of application developers and curious users. It's hard to predict exactly what will come next, but we can say at least that new applications will strive to be social, mobile, analytical, and cloud-based. The next few sections explore some of the most recent technologies being used in GIS.

3.1 | Channels for Public Input

GIS applications allow the public to find their own answers to property valuation questions, including:

- Comparable sales for similar property types
- Potential school districts in which to live
- Which houses are closest to a library or access to public transportation

Computer widgets that allow users to search for answers are exploding in popularity, and can help valuation offices provide answers to common questions regarding properties and the regional characteristics within a particular jurisdiction. Many government agencies are now inviting users to develop new applications using data within their GIS and other open portals. For example, a citizen might use various characteristics within government data to conduct a very specific search, which can then be shared as a new application with all public users who have the same question. Thus, mobile, web-based GIS tools increase citizen engagement and taxpayer interaction.

Another way to increase engagement is through **crowdsourcing**, which is when users provide online feedback and information about current data sets. For example, a commercial property broker could provide information about observed capitalization rates in a commercially zoned property area. Or, a property owner could update their own property record through self-reports and pictures. Crowdsourcing allows anyone to submit potential revisions to data currently on file.

Of course, anytime that an office provides a mechanism for public feedback, they must also establish internal review procedures to ensure that the source of suggested information is verifiable and valid before said information can be added or revised within the data set. GIS desktop measurement tools are important for confirming information about property identification and descriptions provided by taxpayers.

Setting up and monitoring a crowdsourcing effort takes both time and resources, but it can be a very effective tool for gathering new and timely information. Property valuation units need to have adequate resources to verify all incoming questions and information. As technology increases transparency, it adds to the operational duties of the valuation unit.

▸ 3.2 | Here Come the Drones

Drones, also called **unmanned aerial vehicles** (UAVs), are one of the newest technologies to be used in property valuation. Drones can expedite data collection and verification procedures through their ability to quickly identify or confirm property characteristics. Without a doubt, the use of drones for government applications will explode in the next decade. Drones will be used to collect supplemental imagery and could replace aerial imagery. Some experts predict that drones will be able to capture information at the level of detail that a land surveyor is required to record for a land transaction. As drones improve and their capacity increases, the amount of time and money it takes to capture images and measurements will get lower and lower.

Updating the images for property records can be completed on-demand using drones programmed with GIS routing software. This ensures that the most current and timely images are available for developing valuations and defending an appeal.

Developing countries interested in creating land administration systems are piloting programs that use drones to accomplish operational tasks that historically required field personnel to accomplish. As drones and their locational and geographic identification technologies advance, they will record imagery for the following purposes:

- Discovering property
- Defining land boundaries and refining property characteristics
- Observing locational influences on property, such as traffic and other environmental issues

▸ 3.2 | Improved Satellite Imagery

Improvements in satellite technology continue to increase the accuracy of the images captured, while decreasing the physical size of the hardware. More accurate imagery means that it will be easier to decipher a photo's details to use in property measurement and verification. And as satellites become more agile, they will be able to be used more globally, and valuers will have access to new images that can be purposed for valuation of real estate for taxation in ways that were not possible before.

▸ 3.4 | LiDAR Technology

LiDAR is a relatively new surveying method for measuring distance. LiDAR stands for "light detection and ranging," and is a remote-sensing technology that can calculate distance by measuring the reflection of pulsing light (emitted from a laser) on a target surface. For valuation offices, LiDAR can be used to create high-resolution maps with extremely high accuracy. It can also be used to make digital 3-D representation of buildings.

There are all sorts of new gadgets and tools available in the realm of digitized imagery and electronic measurement. And these smart devices will only continue to improve over time, giving valuers more detailed visual information, more accurate property measurements, and more accurate overall data. Office personnel who used to perform data collection functions can then be recast into roles related to public response and data analysis.

3.5 | 3-D Modeling

3-D modeling literally provides a whole new dimension for valuers to work in. GIS imagery often includes a picture of a property as viewed from the outside, but 3-D modeling allows you to add characteristics to a property on a floor-by-floor basis. Creating 3-D models of large condominiums, apartments, offices, or any other buildings that offer the occupants a favorable view, allows for this particular aspect of value to be visualized, creating greater accuracy in determining a building's value with respect to its composition. 3-D models can be viewed from all angles and directions in order to verify favorable views, which is an attribute that has always contributed to value in real estate.

3.6 | Virtual or Holistic GIS

To take 3-D models one step further, GIS programs are developing virtual reality and augmented reality applications. Holographic images can be placed on maps to create virtual buildings, communities, and even cities. Virtual GIS is a technology that is just now emerging and will have a huge impact on mass appraisal modeling by offering this visualization to potential real estate markets, or by offering virtualized data to government agencies where data do not exist. If you can create an image of a property, then you can estimate its measurements, and ultimately estimate its value entirely through virtualization. This technology could assist countries who are trying to develop property tax administration systems in a cost-effective manner when other data are either not available or need to be supplemented.

4 | Open-source software

Open-source software can be defined as software programming or coding that anyone can use, modify, or enhance free of charge. Due to its open nature, this kind of software can change rapidly in response to collaborative input provided by the public. The most common types of open software that are used in valuation offices are statistical software and GIS software. This is a very inexpensive approach to the research and development of new mass appraisal techniques, including the identification of locational influences that should be considered for supplemental use in mass appraisal modeling. More importantly, computer solutions can be developed quickly by leveraging open source software in the cloud that responds rapidly to a user's needs.

4.1 | Advancing spatial decision support systems with GIS technologies

Government valuation offices will see an expanded use of **spatial decision support systems (SDSS)**. Spatial support systems are interactive computer-based systems used alongside GIS to facilitate decision-making. They are sometimes called decision support systems, especially when used for policymaking. SDSS is more specifically being used by valuation offices for refinement of valuation modeling.

▸ 4.2 | Data Repositories

Data repositories can essentially be defined as very user-friendly online data storage destinations. They are sometimes also referred to as **data warehouses**. Professional associations, universities, and governments can share data regarding property characteristics and compare the valuation methods and models they have to determine value. Global data-sharing initiatives are being fostered by IAAO and other international valuation industry associations. Professional associations dedicated to improving land administration and property tax systems worldwide are working together to unify data.

Repositories such as IAAO's Research Exchange (found within IAAO's library) are programmed to be more advanced and interactive in nature, allowing data to be shared with added comments and direction. Data usually require data definition keys, so an author must clearly define the data and material that have been uploaded. Including a definition and allowing for comments and explanations is a much better way to share data than simply sending undefined data sets out to an open portal.

Online repositories allow for the exchange of all types of valuation data, and even facilitate the sharing of new mapping tools, GIS applications, and new ideas regarding spatial modeling and calibration. These core components of value estimation will require continuous refinement as the dynamics of our world change. Data repositories open valuation offices to comparative data and research exchange at an international level, which increases our knowledge base and ability to comprehensively respond to valuation challenges.

▸ 4.3 | Blockchain

Blockchain is a new computer programming method used to record the exchange or transfer of real estate. This cloud-based program is able to record real estate transactions over multiple platforms, rather than being confined to the silo of a single recorder or registry office. Processing real estate transactions electronically is the way of the future, and if it is not a blockchain technology, then another derivation will arise that makes it easier to file and record land registry information.

New technology will transform the amount of information contained in recordings, while increasing the transparency and accuracy of real estate transactions globally. With cheaper and more accurate information available regarding real estate transactions comes more accurate property tax records and value predictions.

▸ 4.4 | Fit-for-Purpose Land Administration

The term "fit-for-purpose" in land administration comes out of recent efforts to develop land administration systems in countries where they do not currently exist. A fit-for-purpose land administration system is one that is designed to meet the needs of the people who reside on the land. Land boundaries are decided upon and drawn on a mapping system with the participation of the people who reside in that region. Land boundaries will be determined based on how land is currently being used, rather than through the traditional (but costlier) methods of surveying or researching to create legal boundaries for official recordings.

The goals of a fit-for-purpose land administration system is to be participatory, affordable, and quick to design and implement. Frankly, they are considered "good enough" to get the land administration system started. The primary goal is to draw simplified land boundaries that meet government needs for economic development and social delivery. This method has been described as a realistic and pragmatic approach to building a system that can then be improved over time.

5 | Overcoming Data Challenges

The greatest obstacle to data sharing is that so many valuation data sets originate at a very local level. Currently, data definitions, data characteristic names, and the interpretation of what each element means vary dramatically across valuation jurisdictions worldwide. What one jurisdiction calls a "basement area," another jurisdiction calls "lower-level."

Commonality in both the definition of data and standards for measurement has not yet been achieved in property tax data sets across the world. The electronic coding schemes used for identifying property in taxation systems also tend to be very different. Efforts are underway to standardize data definitions and measurement techniques for real estate taxation purposes.

5.1 | National Parcel Identification and Data Standardization Initiatives

In the United States, efforts to create a national parcel data identifier for all property throughout the country have been underway for more than two decades. There is also an ongoing effort to create a nationalized geospatial standardized data platform. Advocates of these efforts are trying to create a method for converting the large variety of local property indication numbering systems into a national parcel identifier, and to a standardized property characteristic data set for use in cadastral mapping.

There is consensus that the national parcel identification number should follow the numbering conventions used in the **Public Land Survey System (PLSS)**. As described in Chapter 2, the PLSS is the surveying method developed in the United States to survey and record land dimensions for the creation and updating of plats or parcel maps. The PLSS is often called the **rectangular survey system** and was created by a land ordinance back in 1785. Since then, the PLSS has been the primary survey method used in the United States by a multitude of government agencies at every level. It therefore stands to reason that the principles and measurement methods used in the PLSS were recommended for use in the creation of a national parcel identifier.

Although numerous federal and local agencies are on board to achieve the goal of having one national parcel identifier, it is still not an easy task to accomplish. There are many local government agencies who have created their own derivations and specialized numbering schemes for the identification of the property within their local jurisdictions. If a conversion method is adopted in the future and becomes successful, then the sharing of property information at a national level will finally be possible in the United States. Coordination among government units that deal with land-related issues will improve dramatically.

Many countries already have national property or parcel identification systems since the property tax administration function is carried out on a national level. Countries that administer property tax systems on a local level, or who are developing national property tax systems for the first time, face the same challenges as the United States. Internationally, local differences in data definitions and collected data characteristics also present an issue.

The benefits of national parcel maps are many, and include the following.

- **Identifying a property's ownership and characteristics with ease and accuracy would help the coordination of any national effort.** This is especially important in the case of a natural disaster. National agencies working alongside local agencies would have instant access to accurate and uniform information for emergency responders, resulting in improved reaction time.

- **For valuation offices, a national aggregate parcel database and cadastral map would expedite the use of shared government resources** and offer commonality of data elements among valuation techniques.

- **Valuation offices would become the keepers of an authoritative set of data.** By standardizing parcel data, the hope is there will be reduction in non-authoritative and shadow data, so that government agencies are confident in the data they plan to apply to any situation or global purpose.

- **Having a standardized data set provides valuation offices with comparable property analysis ay a national level.** The advantages of this are many. For example, unique property types are very difficult to find good comparable data for. Properties such as hospitals, quarries, sports facilities, or large manufacturing and industrial properties would benefit from being analyzed and compared to other properties on a national level in order to increase uniformity in applied property valuation methods. Using a national parcel data map would also display the differences in location between properties and assist in creating adjustments based on location, and in choosing the most comparable locations.

- **A national property data set could be combined with other national information and open data sets that concern real estate markets,** including mortgage interest rates, analyses of nationwide economic indicators of growth, national real estate sales activity (identified for every property class and potentially divided by submarkets), foreclosure rates, and any other economic trends that can be identified and that have an influence on property value, either nationally or at a regional level.

5.2 | The Coalition of Geospatial Organizations (COGO)

Neal Carpenter, President and CEO of the Sidwell Company, has served as IAAO's representative to the Coalition of Geospatial Organizations (COGO), of which IAAO is a member. COGO was formed in 2008 and membership is composed of 13 national professional societies, trade associations, and membership organizations in the geospatial field, representing more than 170,000 individual producers and users of geospatial data and technology.

COGO will advocate a position on proposed federal legislation that influences geospatial data, and often provides specific input on legislative details. COGO's collective expertise regarding geospatial policies and the structure of cadastral data sets only speaks out when its member organizations vote unanimously to do so. COGO has been a strong public policy advocate, particularly on behalf of standardizing and nationalizing property-related geospatial data.

COGO has supported the adoption of a national geospatial data framework that is intended to structure data in such a way that those data will be trusted, standardized, and described according to one common standard. Another goal they have is for a national map to be made publicly available at minimal or no cost. Currently there is no national program or funding dedicated to the development and maintenance of parcel data in the United States, but efforts will continue to be made by many organizations to find the funding and programs necessary to create a national spatial data parcel map that utilizes a uniform spatial data infrastructure.

Neal comments, "In America, there are about 150 million federal land parcels that are currently maintained by about 6,700 land record data stewards in over 3,200 counties and equivalent units of local government. We must work to support a comprehensive approach to assembling parcel information from local government stewards, and we do not have a strong program to create and support a national cadastral data theme at this point.

"It should be noted that there is no national coordination of cadastral data in the United States. By contrast, you can take a good look at the 28 countries of the European Union, which have already made a cadastral parcel map for the foundation of the infrastructure for spatial information in the European Community. COGO, particularly with the assistance of IAAO, understands that the framework of data we developed—which acknowledges the absolute importance of local valuation partners—must be adopted. We need a new bottom-up approach, rather than a top-down approach, to bring about the creation of a national cadastral parcel data layer and map. Valuation officials globally agree that local government data are always the best and most authoritative source for parcel data and property characteristic updates.

"Since parcel data is typically created and maintained at a local level, the challenging task for the future is unifying these disparate data sets and making them consistent in order to respond to future business and government needs."

Neal emphasizes, "Your cadastral data will eventually be used as part of a national parcel map. The question is, will local valuation officials have a voice in determining how their maps will be shared and brought into that system? IAAO will continue to be a leading voice in determining how that map will be created, maintained, used, and shared."

Solutions to this standardization problem will be created and shared over the next 10 years, and we might even see a global parcel data map in the next 20 years. Recent web-enabled GIS technologies spur the imagination toward the possibility of creating a global map that is used by all countries and more than 100 million parcels worldwide. The vision and even more ambitious idea of an international parcel identification numbering system is something that will take years to develop globally, and can only happen with the assistance of many valuation organizations and other stakeholders working together to achieve this goal.

▸ 5.3 | The International Property Measurement Standards Coalition (IPMSC)

The International Property Measurement Standards Coalition (IPMSC) is working to create a complete set of standards for the measurement and definition of properties worldwide. IAAO is a member organization of IPMSC. More information about them can be found at **www.ipmsc.org.**

IPMSC is composed of 80 professional and not-for-profit organizations located around the world and is constantly growing. The goal of IPMSC is to ensure that all property is measured in a consistent way—no matter where in the world those measurements are taken—by becoming the primary method of property measurement. They want to create a more transparent global marketplace, develop greater public trust and stronger investor confidence, and ultimately increase market stability worldwide.

The coalition has an independent standard-setting committee (SSC) that includes technical experts from 12 countries who collectively have expertise in 47 different real estate markets. IPMSC's standards define the aim and scope for each standard and detail the measurement calculations that are to be used for every property type. If the entire world could use the same language and definition of building measurements, there would be much greater transparency in global markets and real estate transactions. In addition, comparative analyses could be conducted by members of the valuation industry throughout the world. IPMSC's efforts—supported by IAAO—will continue to expand and be communicated as they pursue their goal of standardized real estate measurements across the world.

6 | Conclusion

In the future, GIS technologies will interlock with other mass appraisal software, making their use almost seamless. Location coordinates will become the building blocks of massive data sets that will be spatially displayed and mapped worldwide. Using cloud-based GIS technologies opens the valuation industry up to so many new and creative ways to discover and analyze data. Data sharing will be an everyday occurrence as the valuation industry works to improve and augment existing data sets. It is our hope that data standardization will soon be achieved.

We hope, too, that this book has given readers a new understanding of how GIS can be used to improve all aspects of valuation. Every valuation office strives to improve existing data, and GIS technologies are the most cost-effective way to do so, giving users the ability to obtain, maintain, refine, verify, and review property tax records. GIS technology allows for existing data sets to be easily augmented with new data from many different sources (open data sources being particularly helpful) in order to enhance the information available to inform the estimation of property value.

Government agencies can also tell their story more transparently by providing creative mapping and interactive GIS tools to the public for individual exploration. All agencies within a property tax and land administration system can now map information relevant to property taxes. This increased interaction does much to foster trust and improve relationships with the public.

As technology develops further, internal operation models will need to be reinvented in order to maintain valuation offices' ability to keep up with changing times. GIS technology can aid in that reinvention as an even greater number of GIS creative solutions and devices continue to emerge over the next 20 years. More will be expected of the assessment industry in the production of value estimates as online access to data and open source software becomes commonplace. Online interactive data centers for public use and increased transparency will be a mainstay of government.

Government valuation leaders will need to be diligent in the promotion of continuing education and lifelong learning for valuation professionals, focusing not only on valuation topics but also on keeping up to date with GIS and all of the new technologies that are transforming how we work. As many older professionals within the valuation industry are heading toward retirement, significant opportunities for professional success will hit a record high for younger professionals. Providing new valuation industry leaders with the training and technology tools they need to succeed should be a universal goal.

IAAO is committed to be on the cutting edge in the development of mass appraisal modeling improvements by using GIS tools to discover new data patterns. GIS enables the creation of more precise location factors that influence value for mass appraisal modeling. IAAO is committed to continuing to:

- Provide targeted training to educate the valuation industry and promote designations that are practical and increase the users' proficiency
- Promote both data sharing and standardization among the assessment industry across the world
- Share GIS innovations for the implementation of cost-effective valuation operations
- Revise and provide practical standards for cadastral mapping and all other operations contained within valuation offices worldwide
- Assist in the development of robust property tax systems across the world

Our appendix includes targeted GIS resources for valuers that can assist government leaders in starting their own GIS technology journey. We hope that it helps you in your quest to reach our shared goals of transparency, uniformity, and equity within valuation offices.

Appendix of GIS Resources

ORGANIZATIONS

The following organizations contain additional information on how to integrate GIS technologies into the tasks carried out by a valuation office.

The International Association of Assessing Officers (IAAO)

- *Main site:* www.iaao.org
- *Valuation standards:* www.iaao.org/standards

Esri

- *Main site:* www.esri.com

Urban and Regional Information Systems Association (URISA)

- *Main site:* www.urisa.org

GLOSSARIES

Diving into the world of GIS requires you to become familiar with an entirely new lexicon. While many of the GIS terms used in this book are defined within the chapters themselves, the glossaries below (organized alphabetically by source) may provide an additional understanding of these terms, and certainly include many terms that were not discussed here. Use them to expand and refine your GIS vocabulary.

- ***ArcGIS, Insights for ArcGIS Glossary:*** http://doc.arcgis.com/en/insights/enterprise/latest/reference/glossary.htm
- ***Bureau of Land Management (BLM), Glossaries of BLM Surveying and Mapping Terms:*** https://www.blm.gov/or/gis/geoscience/files/BLMglossary.pdf
- ***City of Fort Collins, Glossary of GIS Terms:*** https://www.fcgov.com/gis/terms.php
- ***Dalhousie University, GIS Glossary:*** https://libraries.dal.ca/help/gis-glossary.html
- ***Department of Homeland Security, Geographic Information System Software Selection Guide:*** https://www.dhs.gov/sites/default/files/publications/GIS-SG_0713-508.pdf
- ***Esri, GIS Dictionary:*** https://support.esri.com/en/other-resources/gis-dictionary
- ***Federal Geographic Data Committee (FGDC), Lexicon of Geospatial Terminology:*** https://www.fgdc.gov/policyandplanning/a-16/lexicon-of-geospatial-terminology
- ***GISGeography, Geospatial Definition Glossary:*** https://gisgeography.com/gis-dictionary-definition-glossary/
- ***Land Info, GIS Definitions:*** http://www.landinfo.com/resources_dictionary.htm
- ***Open Geospatial Consortium (OGC), Glossary of Terms:*** http://www.opengeospatial.org/ogc/glossary
- ***UC Berkeley, Dictionary of Abbreviations and Acronyms in Geographic Information Systems, Cartography, and Remote Sensing:*** http://www.lib.berkeley.edu/EART/abbrev.html
- ***Wiki, GIS Glossary:*** http://wiki.gis.com/wiki/index.php/GIS_Glossary

ADDITIONAL GIS RESOURCES

The following three lists go one step further, providing additional GIS resources collected by various government and nongovernment agencies.

At the federal level...

- **U.S. Environmental Protection Agency**
  - *Geospatial site:* https://www.epa.gov/geospatial
- **Federal Geographic Data Committee (FGDC)**
  - *Main site:* https://www.fgdc.gov
  - *Geospatial platform:* https://www.geoplatform.gov/
  - *Short glossary:* https://www.geoplatform.gov/glossary-of-terms
- **U.S. Department of Housing and Urban Development**
  - *GIS data services:* https://www.huduser.gov/portal/egis/index.html
- **U.S. Geological Survey (USGS)**
  - *GIS data and tools:* https://www.usgs.gov/products/data-and-tools/gis-data
  - *National map:* https://nationalmap.gov/
- **U.S. Census Bureau**
  - *Topologically Integrated Geographic Encoding and Referencing (TIGER) products:* https://www.census.gov/geo/maps-data/data/tiger.html

At the state level...

- **New Jersey Geographic Information Network:** https://njgin.state.nj.us/NJ_NJGINExplorer/index.jsp
- **Idaho Geospatial Office:** https://gis.idaho.gov/
- **North Carolina Department of Transportation (NCDOT) GIS Services:** https://connect.ncdot.gov/resources/gis/Pages/default.aspx
- **Office of Geospatial Information for Missouri GIS Resources:** https://oa.mo.gov/information-technology-itsd/it-governance/office-geospatial-information

Nongovernment resources...

- **The R Project for Statistical Computing:** https://www.r-project.org/
- **Designing effective maps and cartography**
  - *Designing Better Maps: A Guide for GIS Users, by Cynthia Brewer (2005, Esri Press)*
  - *Designed Maps: A Sourcebook for GIS Users, by Cynthia Brewer (2008, Esri Press)*
  - *Color advice for cartography: http://colorbrewer2.org/*

RECENT PUBLICATIONS

Finally, we offer a curated selection of recent academic publications concerning GIS technologies. Through these papers, you can explore a number of topics in great depth, and gain insight into the ways that GIS is evolving and might be used in the future.

- Andarias, R. (2015). Technology and tax administration: a local tax agency in Spain. In *2015 Proceedings of the 81st Annual International Conference on Assessment Administration.* Kansas City, MO: International Association of Assessing Officers.
- Andarias, R. (2016). Technology and tax administration: the case of a local tax agency in Alicante Province, Spain. *Fair & Equitable*, 14 (3), 3-11.
- Anselin, L. (2005). *Exploring Spatial Data With GeoDa: A Work Book.* Spatial Analysis Laboratory, University of Illinois. Center for Spatially Integrated Social Science.
- Aroul, R.R. and M. Rodriguez. (2017). Obtaining better sustainability data for hedonic analysis. *Journal of Real Estate Literature*, 25 (2), 429-443.
- Bidanset, P.E., J.R. Lombard, P. Davis, M. McCord, and W.J. McCluskey. (2017). Further Evaluating the Impact of Kernel and Bandwidth Specifications of Geographically Weighted Regression on the Equity and Uniformity of Mass Appraisal Models. In *Advances in Automated Valuation Modeling* (pp. 191–199). Springer International Publishing.
- Bidanset, P., M. McCord, J. Lombard, P. Davis, and W.J. McCluskey. (2017). Accounting for Locational, Temporal, and Physical Similarity of Residential Sales in Mass Appraisal Modeling: Introducing the Development and Application of Geographically, Temporally, and Characteristically Weighted Regression (GTCWR). *Journal of Property Tax Assessment & Administration.* 14(2), 5.
- Bidanset, P., D. Fasteen, and B. Jones. (2016). GIS in the assessor's office: a study of emerging trends. In *2016 Proceedings of the 81st Annual International Conference on Assessment Administration.* Kansas City, MO: International Association of Assessing Officers.
- Birch, J.W., M.A. Sunderman, and E. Radetskiy. (2017). Reducing Vertical and Horizontal Inequity in Property Tax Assessments. *Journal of Property Tax Assessment & Administration*, 14(2).
- Borst, R.A. (2014). "Improving Mass Appraisal Valuation Models Using Spatio-Temporal Methods." *International Property Tax Institute*, Toronto, Ontario, Canada.
- Brotman, L., W. Robertson, and K. Jones. (2015). State tax agencies assist counties with parcel mapping. *Fair & Equitable*, 13 (10), 3-8.
- Burle, B. and P. O'Connor. (2017). Finding statewide comparable sales using GIS. *Fair & Equitable*, 15 (7), 32–33.
- Crane, E., et al. (2015). Technology: realizing the benefits and managing the challenges, part 1. In *2015 Proceedings of the 81st Annual International Conference on Assessment Administration.* Kansas City, MO: International Association of Assessing Officers.
- Demetriou, D. (2016). GIS-based automated valuation models (AVMs) for land consolidation schemes. In *Proceedings from the 6th International conference on Cartography and GIS*; June 13-17, 2016 in Albena, Bulgaria.

RECENT PUBLICATIONS *(cont.)*

- d'Amato, M. and T. Kauko. (2017). Appraisal Methods and the Non-Agency Mortgage Crisis. In *Advances in Automated Valuation Modeling* (pp. 23–32). Springer, Cham.
- Fasteen, D. (2016). Demystifying GIS Technology for Assessment Professionals: Insights, Trends, and Examples of GIS in the Property Assessment Industry. In *2016 International Research Symposium proceedings*. Kansas City, MO: International Association of Assessing Officers.
- Fasteen, D. (2016). Factors that influence the adoption of geographic information systems in a professional work environment: a study of the property assessment profession. Grand Forks, ND: University of North Dakota.
- Hallewa, A. (2016). Mass appraisal techniques combined with GIS - ownership transfer project in the Israel Land Authority. In *FIG Working week 2016 proceedings.*
- Lomax, M.J. (2015). Transforming the real estate industry and valuation using technology. *Canadian Property Valuation*, 59 (3), 24-29.
- Nguyen, H., R. Guirdry, and L. Hara. (2015). Collecting field data with an iPad. In 2015 *Proceedings of the 81st Annual International Conference on Assessment Administration.* Kansas City, MO: International Association of Assessing Officers.
- Nunlist, T. (2017). Virtual valuation: GIS-assisted mass appraisal in Shenzhen. Land Lines, 29 (4), 8–13.
- Oud, D.A.J. (2017). GIS based property valuation: objectifying the value of view. Utrecht, Netherlands: Utrecht University.
- Paskevicience, A. (2015). GIS application for mass valuation. In *Case studies and supporting materials of the 2015 conference on property valuation and taxation in Europe and Central Asia.*
- Scarborough, K. (2016). GIS speeds parcel editing and map deployment in Osceola County, Florida. *Fair & Equitable*, 14 (2), 25–26.
- Tomlinson, R. (2013). *Thinking about GIS: geographic information system planning for managers.* Redlands, CA: Esri
- Wade, T. and S. Sommer. (2006) *A to Z GIS: an illustrated dictionary of geographic information systems.* Redlands, CA: Esri
- Welcome, P., et al. (2015). Defending commercial appraisals with location analytics. In *2015 Proceedings of the 81st Annual International Conference on Assessment Administration.* Kansas City, MO: International Association of Assessing Officers.
- Williamson, I., Enemark, S., and et al. (2010). *Land administration for sustainable development.* Redlands, CA: Esri

About the Authors

Margie M. Cusack serves as the Research Manager for the International Association of Assessing Officers, creating original research and publications studying emerging trends and best practices for application in assessment administration. Margie brings real-world experience in property tax administration to IAAO research. Prior to this position, Margie was the Director of Real Estate for the Cook County Clerk's Office. Margie was responsible for calculation of approximately $14 billion in local taxes derived from approximately 1.86 million properties. Margie's thirty years of property tax administration specialized in GIS system expansion, property valuation and tax policy development. Prior to leading the Clerk's Office Taxation Unit, Margie was the Chief of Valuations at the Cook County Assessor's Office. Margie was a lead in creating an integrated GIS system that is transferable to use at any government level. She holds an Bachelor's degree from University of Wisconsin–Madison and two Master's degrees from the University of Chicago, Harris School of Public Policy and the Department of Political Science. Margie has also taught Public Policy and Administration at the University of Illinois–Chicago.

Paul E. Bidanset currently serves as the IAAO's valuation research project manager. He is a Ph.D. candidate at Ulster University in Belfast, Northern Ireland, where he has developed a new mass appraisal model under the supervision of doctors Billy McCluskey, Peadar Davis, and Michael McCord. Prior to working for the IAAO, Paul was the computer-assisted mass appraisal (CAMA) modeler for Norfolk, Virginia. His expertise primarily focuses on using GIS and statistical modeling to improve valuation accuracy, uniformity, and defensibility for assessment offices. He holds both a bachelor's of arts and master's of arts in economics from James Madison University and Old Dominion University, respectively.

Daniel J. Fasteen, Ph.D. is the Director of Special Projects for BIS Consulting (Farmers Branch, TX) where he designs and implements analytical services and solutions for mass appraisal. Daniel has been in the property valuation profession since 2010 working for local government jurisdictions in North Dakota and Minnesota. His primary areas of expertise include the design and application of spatial automated valuation models (AVMs) to mass valuation problems, and the integration of GIS and CAMA systems technology. Daniel holds a Master's degree in Geography and a Ph.D. in Research Methodologies and Applied Statistics, both from the University of North Dakota. He has presented papers at numerous conferences and international symposiums, currently serves on the editorial board of the Journal of Property Tax Assessment & Administration, and is a former Chair of the IAAO Research Subcommittee.

Made in the USA
Columbia, SC
09 September 2019